Katie

Broad Leys Publishing Ltd

**Organic Poultry**

First published by Broad Leys Publishing Ltd: 2005



A catalogue record for this book is available from the British Library.

ISBN: 0 906137 36 5

Outside front cover: Lohmann Tradition layers

Outside back cover: Top - Pekin cross table ducks
Middle: Bronze turkey
Bottom: Embden cross geese

Unless indicated otherwise all photographs were taken by the author.

For details of other publications please see Page 120.

Broad Leys Publishing Ltd
1 Tenterfields,
Newport, Saffron Walden,
Essex CB11 3UW, UK.

Tel/Fax: 01799 541065
E-mail: kdthear@btinternet.com
Website: www.blpbooks.co.uk

# Contents

# Preface

*The trouble is there's organic and organic!*
(Supermarket shopper. 2005)

In recent years there has been an increasing demand for wholesome food that has also been produced humanely. Free-range eggs spearheaded the change, followed by the rapid development of non-intensive broiler birds. Britain led the way in free-range egg production, while France, with its traditional emphasis on gastronomic excellence, has shown how traditional breeds and rearing methods are appropriate for free-range table bird production. It is not just a question of keeping birds on free-range, however. The housing, choice of breeds, diet, flock intensity and other details of integral management are just as important. It is with these specifics that organic production is superior to other methods.

This book is for those who are thinking of keeping organic poultry for the sale of eggs or table birds. It is also applicable to those who just wish to keep small household flocks in the best way possible.

It does not aim to cover every detail of the organic standards, for this would require several tomes. Standards are also subject to change, with new interpretations appearing at regular intervals. Costings are also deliberately omitted for these quickly become out of date.

What the book does set out to do is to give an overview of the basic UK Organic Standards and then indicate where there might be variations in other standards. It does not purport to be comprehensive, but details of where to obtain further information are provided for the prospective producer.

*The Soil Association*, in its 2004 Report, commented that the problems surrounding different standards fosters confusion amongst consumers over differences between organic and free-range, as well as within the organic sector itself. I hope therefore, that this book will help to clarify the current situation, as well as to reduce some of the confusion that currently reigns amongst both consumers and prospective producers.

The emphasis of the book, however, is on the practicalities of keeping small flocks. In this respect, I have drawn heavily on my own experiences and on those who are active in the field, as producers or researchers. I have also tried to draw together the latest research findings in relation to free-range and organic poultry, and references to these will be found in the text.

I am grateful to the many individuals and organisations who have contributed help, advice and information, and I am particularly grateful to the producers who welcomed me to their farms. *(Katie Thear, Newport, 2005).*

# What is Organic?

*Effective communication demands that you define your terms.*
(Miss Metcalfe, the author's old English teacher)

My old English teacher always stressed the importance of defining terms, otherwise, she would ask us gravely, how would anyone know what anyone else was talking about? So, it is prudent to follow her sterling example and start with some definitions.

### Poultry

Officially, the following birds are recognised as poultry and are therefore subject to regulation: chickens, ducks, geese, turkeys and guinea fowl.

### Organic

There are three ways of describing what is meant by the term organic: the dictionary definition, the traditional definition as most people know it, and the legal one.

**Dictionary definition** The dictionary definition is that it is a coordinated whole with various factors contributing to an organised body, where the connected and interdependent parts share a common life.

**Traditional definition** Those who have been organic before there was even a legal definition of the term, understand that their activities are benign, humane and enhance the environment rather than damage it. They work on the following principles that have been defined by the International Federation of Organic Agriculture Movements (IFOAM):

- Produce food of high quality in sufficient quantity.
- Interact in a constructive and life-enhancing way with natural systems and cycles.
- Consider the wider social and ecological impact of the organic production processing system.
- Encourage and enhance biological cycles within the farming system involving micro-organisms, soil flora, plants and animals.
- Maintain and increase the long-term fertility of soil.
- Maintain the genetic diversity of the production system and its surroundings, including the protection of plant and wildlife habitats.
- Use as far as possible renewable resources in locally organised production systems.

• Create a harmonious balance between crop production and animal husbandry.
• Give all livestock conditions of life with due consideration for the basic aspects of their innate behaviour.
• Minimise all forms of pollution.
• Process organic products using renewable resources.
• Allow those involved in organic production and processing, quality of life which meets their basic needs and allows an adequate return and satisfaction from their work, including a safe working environment
• Progress towards an entire production, processing and distribution chain that is socially just and ecologically responsible.

**Legal definition** When it comes to the production of food, the word organic also has a legal definition so that consumers may be assured that so-called organic produce has been produced in a humane way and by sustainable management that does not damage the environment. It applies to horticultural and farming methods, crops, foods, animals and poultry and there are two sets of regulations and standards that apply:

• *EU Regulation EC2092/91* which has the status of law throughout the European Union. This sets the minimum standards that are required for organic production.

• *Interpretation of Standards* within the UK. The EU Regulations specify that each member country has its own interpreting organisation. In the UK, this is the Advisory Committee on Organic Standards (ACOS). It was originally called United Kingdom Register of Organic Food Standards (UKROFS), and is part of the Department of Food and Rural Affairs (DEFRA). The interpreted standards are referred to as the UK Organic Standards.

### Function of ACOS

ACOS (originally UKROFS) interprets the European Union legislation for organic production within the UK. It advises the Government on organic matters and is responsible for research and development in the field. It also ensures that producers are certified as organic producers and are subject to inspection. However, the actual jobs of certification and inspection are put out to various organisations who must themselves be registered with ACOS.

## Registered certification bodies

The registered bodies can set their own standards for their members. These standards may be higher than those officially required but they must not fall below those of the UK Organic Standards.

There are several approved certification organisations. Each one has its own identifying code number that must appear on the labelling of its members' produce. They also have a logo that only registered producers may use on their packaging.

**Soil Association Certification Ltd** (Code: UK5)
The Soil Association has been promoting organic farming and horticulture since 1946, long before there was a legal definition of organic. Soil Association Certification Ltd was set up in 1973 and is a wholly owned subsidiary of the association. It has its own standards which are higher than those of the UK Organic Standards.

**Organic Farmers and Growers Ltd** (Code: UK2)
Organic Farmers and Growers started as a marketing cooperative in the 1970s, later becoming a certification body when organic farming became legally defined and controlled. OF&G Standards generally conform with those of UK Organic Standards.

**Organic Food Federation** (Code: UK4)
Apart from a few points, the OFF Standards conform with those of the UK Organic Standards.

**Organic Trust Ltd** (Code: UK9)
The standards generally conform with those of the UK Organic Standards.

**Biodynamic Agricultural Association/Demeter** (Code: UK6)
The standards conform with those of UK Organic Standards with additional requirements based on Rudolph Steiner's biodynamic principles.

**Scottish Organic Producers' Association** (Code: UK3)
The standards generally conform with those of UK Organic Standards.

**Irish Organic Farmers' and Growers' Association** (Code: UK7)
The standards vary slightly from those of the UK Organic Standards.

**Quality Welsh Food Certification Ltd** (Code: UK13)
The standards generally conform with those of UK Organic Standards.

**Ascisco Ltd** (Code: UK15)
This was set up by the Soil Association in 2003 for those who, while able to meet the minimum UK Organic Standards, are not in a position to meet the full Soil Association standards.

**CMi Certification** (Code: UK10)
This company provides certification for food processors and suppliers of organic produce.

## Other standards

There are several other sets of standards which are not necessarily organic but which nevertheless are relevant. Most of their recommendations are already covered by the organic standards.

### Farm Animal Welfare Council Recommendations (FAWC)

This is an independent advisory organisation that was set up by the Government in 1979. It keeps the welfare of livestock under review and advises on changes that may be necessary. It emphasises the five freedoms that are essential:

- **Freedom from hunger and thirst** - by ready access to fresh water and a diet to maintain full health and vigour.
- **Freedom from discomfort** - by providing an appropriate environment including shelter and a comfortable resting area.
- **Freedom from pain, injury or disease** - by prevention or rapid diagnosis and treatment.
- **Freedom to express normal behaviour** - by providing sufficient space, proper facilities and company of the animal's own kind.
- **Freedom from fear and distress** - by ensuring conditions and treatment which avoid mental suffering.

## Code of Recommendations for the Welfare of Livestock

DEFRA have produced a range of free advisory publications for poultry and livestock. They are available from DEFRA but are not organic in coverage.

## Freedom Food

This is a subsidiary of The Royal Society for the Prevention of Cruelty to Animals (RSPCA). It was set up to provide minimum welfare standards, but these do not specify free-range or organic as necessities. Producers who are registered with them can use their logo on packaging for their produce. Some producers are registered as organic and Freedom Food and use both logos on their produce.

## Free-range

Free range is a Special Marketing Term (SMT) that was originally defined by the Egg Marketing Regulations. Eggs described as 'free range' must be produced by hens that are kept at a maximum of 2,500 birds per hectare (2.47 acres) and have continuous daytime access to open air runs. The regulations also cover details of housing, size of pop-holes, and so on.

All organic poultry must be on free-range, but all free-range poultry do not necessarily conform to organic standards.

The organic standards include most of the requirements specified by free range regulations, as well as the conditions that are specific to organic management. However, there is variation between the different certification bodies. For example, the basic standards that are followed by most, state that birds should have access to range when weather conditions are suitable and for at least one third of their lives. The Soil Association standards are more stringent. They require birds to have easy access to range during daylight hours, and for the minimum periods indicated below:

Layers: all their laying lives
Table chickens: two-thirds of their lives
Ducks: two-thirds of their lives
Geese: two-thirds of their lives
Turkeys: two-thirds of their lives
Guinea Fowl: two-thirds of their lives.

## Which standards to follow?

The prospective producer will have to decide which of the organisations to register with, and it is a matter of personal choice. The Soil Association standards are the oldest and most widely recognised and many producers are attracted by the fact that, while listing all the UK Organic requirements, they also recommend principles that are more in keeping with the traditional small scale aspects of poultry keeping. For example, they recommend small housing units with a maximum number of birds per house, as follows:

Layers: 500
Ducks: 500
Table birds: 500
Geese: 250
Turkeys: 250
Guinea Fowl: 500

## Comparison of Standards for Laying Hens

*Basic Free range*

| | |
|---|---|
| Stock density outside | 2,500 birds per hectare (1 per 4 sq.m) |
| Pasture | Mainly covered with vegetation. |
| Outside shelter | Required. |
| Type of house | Fixed or moveable. |
| Stock density inside | 9 birds per square metre. |
| Maximum in house | 3,000 birds. |
| Popholes | 2m x 45cm. 1 per 600 birds. |
| Nest boxes | 8 birds per nest box. |
| Perch | 15cm per bird. |
| Feeders | Track: 10cm per bird. Circular: 4cm per bird. |
| Drinkers | Bell: 1 per 100 birds. Nipple: 1 per 10 birds. |

*Basic UK Organic*

| | |
|---|---|
| Stock density outside | 2,500 birds per hectare |
| Pasture | Grass sward to be maintained. |
| Outside shelter | Required. |
| Type of house | Fixed or moveable. |
| Stock density inside | 6 birds per square metre. |
| Maximum in house | 3,000 birds. |
| Popholes | 2m x 45cm. 1 per 500 birds. |
| Nest boxes | 8 birds per nest box. |
| Perch | 18cm per bird. |
| Feeders | Track: 10cm per bird. Circular: 4cm per bird. |
| Drinkers | Bell: 1 per 80 birds. Nipple: 1 per 10 birds. Outside feeders and drinkers. |

*Soil Association Organic*

| | |
|---|---|
| Stock density outside | 1,000 birds per hectare. |
| Pasture | Well covered with grass. Grass/clover leys, companion grazing with sheep and natural dustbathing areas recommended. Rested for 9 months after each batch of poultry. |
| Outside shelter | Required. Access to woodland recommended. |
| Type of house | Fixed or moveable but the latter preferred. |
| Stock density inside | 6 birds per square metre. |
| Maximum in house | 500 but 2,000 allowed if conditions are suitable. |
| Popholes | 2m x 45cm. Located on different sides of house. |
| Nest boxes | 6 birds per nest box. |
| Perch | 18cm per bird. |
| Feeders and drinkers | Number and distribution adequate to allow development of social groups within the unit. Outside feeders and drinkers. |

However, under certain conditions, the following are allowed by the Soil Association:

| Layers: 2,000 | Ducks: 1,000 | Table birds: 1,000 |
| --- | --- | --- |
| Geese: 1,000 | Turkeys: 1,000 | Guinea Fowl: 1,000 |

This contrasts with the UK Organic Standards that require the following maximum per house, and which many of the other bodies follow:

| Layers: 3,000 | Ducks: 4,000 | Table birds: 4,800 |
| --- | --- | --- |
| Geese: 2,500 | Turkeys: 2,500 | Guinea Fowl: 5,200 |

It should be added that under the UK Standards the latter flock densities are only allowed in houses that were built before 24 August 2005, and for producers who were registered before that date. The derogation was to have come to an end on the above date, but as it was felt that UK producers would be at a disadvantage when compared with producers in other EU Member States, DEFRA has extended the derogation date to 31 December 2010.

My own view, is that *all* the existing standards fall far below the ideal. Research indicates that chickens recognise each other on the basis of their head shape, and are capable of remembering 50-60 other birds in their own flock. (Fölsch. 1996. Quoted in *Organic Poultry Production*. Nicolas Lampkin. 1997). This supports the traditional practice of having a maximum flock size of 50. When you realise that some organic flocks include thousands of birds, it's not surprising that there are widespread problems of aggression. Some organisations also allow beak trimming if there is no other solution, although the Soil Association bans it. However, even their standards allow large numbers. Economic considerations have taken precedence over innate flock needs.

### The small local producer

Those with very small flocks may not be able to justify the cost of organic registration which can amount to several hundred pounds. Even if they are providing conditions that are far better than any of the recognised organic standards, they will not be able to sell their produce as 'organic'.

The Soil Association has a scheme for several producers sharing the costs, if they are within the same area. Alternatively, the produce can be sold at the farmgate and simply not described as organic. A notice could be displayed to the effect that organic feeds are fed to the poultry and that they are kept humanely and allowed to roam outside on pasture. It will then be apparent to callers that the produce is organic in all but name.

The Wholesome Food Association is an organisation that was launched for the benefit of people who are only operating on a small scale and selling produce locally. Their principles are equivalent to, if not higher than any official organic organisation, and all members are expected to share them. There is no system of inspection, however, for as everyone is operating on a local basis, buyers who call to buy produce would soon detect if there were abuses. The association has branches in most areas of the country.

# Land conversion

*That land is the most useful which is the most healthy.*
(*Rerum Rusticarum*. Varro. AD46)

Organic production is a land-based system and all poultry are required to have access to free-range conditions. Land includes pasture and any fields that are used for growing cereal or other crops for their feeds. Such land needs a good level of soil fertility that has been built up by natural means such as the cultivation of legumes, green manures and deep-rooting plants in appropriate rotations. Fertility can be maintained and increased by the use of manure from organic livestock and other organic material that is acceptable to the standards.

Not all the land on a holding needs to be converted at the same time, as long as the conversion area is big enough to be a viable and sustainable unit. This includes having enough pasture area to allow for a regular period of resting after it has been in use. For example, Soil Association standards require that the land is rested for nine months after being used by a batch of laying hens.

An organic area that is in conversion needs to be physically separated from a non-organic area, with stock-proof fences or hedges. Separate records and accounts are also required for the organic area.

### Prohibited inputs

Substances that are prohibited during and after a conversion period include: chemical herbicides, synthetic fertilisers and synthetic pesticides.

Poultry should not be subjected to organo-phosphate treatments, routine and unauthorised veterinary treatments, and the unauthorised use of non-organic foodstuffs as well as genetically modified ingredients.

### Conversion period

If land has not previously been used for organic production, a conversion period of two years is required to give time for a viable and integrated system to be built up. However, this may be reduced to one year if evidence can be produced to indicate that no prohibited substances have been used for the previous twelve months.

The prospective organic producer must be registered with a recognised certification body. A derogation must then be authorised by this inspecting authority so that the land can be monitored during the conversion period. The monitoring involves at least one inspection.

A period of conversion is required before land can be used for organic production.

## The conversion plan

When application is made for land to be converted to organic status, the certification body will require a conversion plan. This details the existing status of the land, how it has been treated in the past and the steps that are to be taken to improve and maintain it. For example, areas of pasture may have relied previously on artificial sources of nitrogen, but to meet organic standards will now need to build up fertility by more traditional methods such as crop rotations and the use of composted material. Details of the proposed rotations and inputs will be required, as well as an overall management plan.

## Organic grants

To apply for financial aid with organic conversion or subsequent maintenance of organic land, DEFRA has an *Organic Entry Level Stewardship* scheme (OELS). This replaces the earlier *Organic Farming Scheme* (OFS).

Potential producers must be registered with an organic certification body and the land must either be recognised as being in conversion or be fully converted already. In the latter case, application is only open to those who have not previously been getting organic funding through a scheme such as the OFS. There is no mimimum holding size.

A Bovans Nera organic egg layer out on pasture to which she must have access during daylight hours. This one is from a Soil Association registered flock.

# Pasture

*Less food is needed by fowls in summer, especially if they are on a good range.*
(*Poultry Keeping and How to Make it Pay*. F.E. Wilson. 1910).

Good pasture is vital to the well-being of poultry flocks. It provides them with a place in which to range, scratch about and exercise, as well as an area in which they can feed on grass tips, clovers, herbs and small invertebrates. Chickens can find suitable spots in which to take take dustbaths while poultry generally can follow their instinctive patterns of behaviour.

All poultry will feed on the fine tips of grasses, while grass is the mainstay of the goose's diet. Sugars and proteins are at their highest levels in springtime grasses. Young grasses are particularly beneficial, for their nutrient value decreases as they mature.

## Nature of the land

The best land for poultry is that which is well-drained yet fertile so that there are no boggy areas to harbour disease but there is sufficient humus in the soil to feed and maintain the pasture. Chalky and sandy soils are free-draining but may be 'hungry' in terms of fertility, while clay soils are prone to waterlogging and 'locking up' of the soil nutrients. Land that is slightly sloping will obviously tend to be better drained than that which is flat.

Liming has the effect of flocculating clay particles so that drainage is improved and the nutrients are released in heavy soils. If the soil is acidic, the lime also increases alkalinity. Ideally, pasture land should have a pH value of 6.5. This can be checked by means of soil tests and action taken as necessary. Agricultural soil testing kits are available from specialist suppliers, or a contractor can be employed to do it. The addition of compost and manure adds bulk and increases the fertility of all soils. Where there is a hard 'pan' or compacted layer on the surface of the soil, ploughing will break it up and so improve the drainage.

## What is pasture?

Pasture is not any old piece of land, just as vegetation is not necessarily grass. In fact, the basic UK Organic requirements that open-air runs be 'mainly covered with vegetation' could arguably cover a multitude of sins. They do state however, that the fertility and biological activity of the soil must be maintained and increased by the cultivation of legumes and green manures in a 'multi-annual rotation programme'.

Pasture may be *permanent grassland* in that it is never ploughed but which over the years has maintained a balance of grasses, clovers and herbs. Such pasture is often found on traditional dairy farms, and is made available to livestock in rotation, or is used for hay. Where it has been well maintained, it represents a valuable asset in terms of conservation meadowland, but if it has been neglected, it may be waterlogged as a result of surface panning or compaction. Pernicious weeds such as thistles, bracken and ragwort may also have become established.

More commonly, pasture is made up of *leys* which provide temporary areas of grass and clover within a rotation system. For example, it may last for three to five years before being ploughed for another crop. The advantage of using leys is that short-growing grasses that are suitable for poultry can be sown, rather than the long stalked grasses that are usually found in permanent pasture, and which are more appropriate for ruminants and for hay crops. The clovers that are included in grass leys can 'fix' atmospheric nitrogen into the soil by the action of their root nodules, so that the overall fertility of the soil is increased.

## Crop rotations

Rotating crops on an area of land does several things. It helps to increase organic matter in the soil and build up its overall fertility. Weeds are kept under control, while the life cycle of parasitic pests is disrupted so that clean land is available for the poultry.

There are many different rotations and there is no such thing as a blueprint. It depends on the individual farm enterprise, geographical location and soil type. Some crops such as wheat and barley are more suited to dry areas, while oats are better in damper areas of the country. Some root crops, such as potatoes are better in heavy, neutral soils, while brassicas such as kale and cabbage, do better in alkaline soils.

Whatever the situation, the basic priciples of rotation are as follows:

- Grow crops suitable to the terrain and climate.
- Deep-rooting crops, eg. roots, should follow shallow ones, eg, cereals.
- Nitrogen-fixing crops, eg, legumes or clovers, should precede those that are heavy consumers of nitrogen, eg, potatoes or leafy crops such as brassicas.
- Never leave the soil bare but maintain the soil cover by means of green manuring with a cover crop that can then be dug in to increase organic matter in the soil and hence the soil fertility.

An example of a crop rotation follows, but it must be emphasised that this is one of many possibilities, and it should not be taken as a definitive blueprint for all situations.

**One example of a Crop Rotation**

| | |
|---|---|
| Year 1 Grass and clover ley<br>Year 2 Grass and clover ley<br>Year 3 Grass and clover ley | Pasture fields made available in turn to poultry. Deep-rooting clover fixes atmospheric nitrogen in soil. |
| Year 4 Cereals (wheat, barley or oats) | Shallow rooted crops. Take advantage of nitrogen provided by clovers. |
| Year 5 Roots (potatoes) | Deep-rooted and help to break up soil for next crop. Alternatively, legumes (or other cover crop) to provide organic matter and nitrogen in the soil. |
| Year 6 Cereals (wheat, barley or oats) | Shallow-rooted crops take advantage of fertility from previous crop. Cereal can be under-sown with a ley mixture so that after cutting, the grass which has had the benefit of initial protection, grows on to provide the next 3-year pasture. |

Other examples are:

Grass/clover ley (3 years) - potatoes - legumes - brassicas - green manure - roots.
Grass/clover ley (2 years) - winter wheat - winter barley - green manure - spring oats

## Green manuring

Green manuring is the practice of sowing an annual cover crop on an area vacated by the previous crop, so that weeds do not colonise it. It is also a way of avoiding soil erosion and adding organic bulk to the soil, thus increasing fertility and preparing ground for future crops.

Beneficial biological activity is increased in the soil and the practice breaks pest and disease cycles. Green manuring can be part of a crop rotation, and is ideal for improving land. Leguminous plants fix atmospheric nitrogen in the soil by the action of bacteria in their root nodules, while others have the effect of adding minerals and humus to the soil. Cover crops can be sown in the spring and dug over in the autumn, while hardy plants can be sown in autumn for over-wintering before being incorporated in the spring.

Once the crops are fully established, they are ready to be dug in, but this should be before flowering takes place so that their seeds do not fall on ground that is earmarked for the next crop in the rotation. However, some plants produce flowers that not only enhance the environment, but also provide habitats and fodder plants for a range of insects, including bees, butterflies and hoverflies whose larvae feed on aphids. Some producers therefore, leave unploughed strips at the edge of fields where the flowers such as Clovers and Phacelia can bloom. Other plants, including herbs may also be included in these areas, not as part of a rotation, but for the provision of self-medication plants for free-ranging poultry. (See Page 20).

### Green Manure Plants

| | | |
|---|---|---|
| Lucerne (alfalfa) | *Medicago sativa* | Not suitable for acid soils. |
| Annual lupins | *Lupinus angustifolia* | Fixes nitrogen. Likes acid soil. |
| Tares (vetch) | *Vicia sativa* | Fixes nitrogen. |
| Red clover | *Trifolium pratense* | Fixes nitrogen. |
| Peas | *Pisum sativum* | Fixes nitrogen |
| Field beans | *Vicia faba* | Fixes nitrogen. |
| Buckwheat | *Fagopyrus esculentum* | Not hardy but good on poor soils. |
| Mustard | *Sinapsis alba* | Quick-growing. Reduces eelworm but can carry clubroot. |
| Phacelia | *Phacelia tanacetifolia* | Quick grower. |
| Hungarian Rye | *Secale cereale* | Good weed suppressant. |
| Italian ryegrass | *Lolium multiflorium* | Very hardy. |

## Pasture plants

Ley mixtures contain grasses and clovers in various varieties and ratios, depending on the pasture required.

### Grasses

Perennial ryegrass makes up the bulk of  grass mixtures. It is hard-wearing and productive, and starts to grow early in the year.

Rough stalked meadow grass produces a thick sward but does better in damp soils. For poultry, which thrive best on well-drained soils, the smooth stalked meadow grass  is better suited to dry conditions. It also does well in shady areas, and as poultry will often range where there is tree cover, it is appropriate in this situation. Timothy is a grass that is good for late season growth and does well in fertile soils. On poorer soils, Meadow fescue might be more appropriate. Similarly, the Brown top bent grass adapts well to poorer soils and is hardy. Crested dogstail is common in all grassland. It does well in all soils except sandy ones and is a rather rank self-seeder but is valuable in winter sheep pastures where sheep and poultry may be sharing pasture.

There are many varieties of these grasses which have been developed for agricultural use in recent years. As referred to earlier, the short-growing ones with good tillering (production of new shoots) are the best for poultry.

### Clovers

Clovers in grass ley mixtures are important in boosting the fertility of soil. Like legumes, they can 'fix' nitrogen from the air, using bacteria in their root nodules. The nitrogen is then available to crops that follow on in the rotation. As with the grasses, clovers also provide a source of protein for the ranging poultry. They are also important as bee fodder plants, with the red clover being particularly visited by bumble bees, while the white is crucial for the honey bees.

A field of legumes grown as part of a crop rotation to add bulk and fertility to the soil.

The red clover, *Trifolium pratense,* is fairly upright in growth but is shorter lived than the white clover, *Trifolium repens,* which is more creeping in its growth. The former tend to be used in short lasting leys or as green manure plants, while the latter are more often found in leys that are kept for a longer period. The white clover's creeping growth also makes it more effective in keeping out weeds and unproductive grasses.

Both clovers generally react well to grazing and mowing, although the white is more vigorous. There are several groups of red clover, including broad-leaved and late flowering varieties. White clovers include wild white, white and asilke. Again, there are many varieties of these that have been developed for agricultural use.

## Ley mixtures

There are many ley mixtures available but until recently, there were few that catered specifically for free-range poultry. A traditional mix for this purpose is indicated below:

| | |
|---|---|
| Perennial ryegrass | 16.0lb (7.25kg) |
| Smooth stalked meadow grass | 11.0lb (4.98kg) |
| Wild white clover | 0.5lb (0.23kg) |
| Crested dogstail | 3.0lb (1.36kg) |
| *Total:* | 30.5lb per acre (13.83kg per 0.4 hectare) |

(Ref: A.K. Speirs Alexander. *Hens on the Land.* 1948)

In recent times, a DEFRA funded project at ADAS Gleadthorpe has used the following mixture:

| | | |
|---|---|---|
| Perennial ryegrass | 31.50 kg | |
| Smooth stalked meadow grass | 4.50 kg | (Ref: *DEFRA/ADAS Poultry Pasture Project.* Gleadthorpe 2004) |
| Brown top bent | 2.25 kg | |
| Timothy | 2.25 kg | |
| White clover | 4.50 kg | |
| *Total:* | 45.00 kg per hectare (2.47 acres) | |

*MAS Seeds*, which was one of the first companies to offer poultry pasture mixes, has the following:

| | | | |
|---|---|---|---|
| Perennial Ryegrass | *Vincent* | 3kg | |
| Perennial Ryegrass | *Lasso* | 3kg | |
| Perennial Ryegrass | *Pagode* | 3kg | (Ref: *MAS Seeds Catalogue*. 2005) |
| Creeping Red Fescue | *Pernille* | 3kg | |
| White Clover | *Alberta* | 0.5kg | |
| Wild White Clover | *Nanouk* | 0.1kg | |
| *Total:* | | 12.6kg per acre (0.4 hectare) | |

In the USA, leys based on lucerne are popular because of the high protein content of these plants. One project also found that clover and alfalfa (lucerne) pasture produced 18% more Omega-3 fatty acids and vitamins. (Ref: *Pasteurized Poultry.* Karsten, Patterson & Crews. Penn. State University. May 2003).

## Herbs

Herbs are increasingly being recognised as playing an important role in the maintenance of poultry health, although little research has been conducted in this area. What is apparent however, is that there is a degree of self-medication in the free-ranging activities of poultry. (Ref: *Intake of Nutrients from Pasture by Poultry*. Walker & Gordon. ADAS. Gleadthorpe. July 2002.) This has always been known traditionally: as long ago as the 1930s it was claimed that poultry exhibited self-selection in their choice of foodstuffs and plants. (Funk 1932 and Graham 1932).

Some culinary herbs, such as sage, thyme, marjoram and rosemary are known antioxidants and may also have antimicrobial properties. That may be so, but it is not always appropriate to include these in with the ley mixture when it is sown. What is possible, however, is to sow herbal strips, particularly at the edge of woodland or by hedgerows, both areas that approximate to the natural habitat of the domestic fowl's ancestor, the Jungle Fowl.

Some of the plants that are appropriate in these areas include the following, but please be aware that some are invasive and are regarded as nuisance weeds so you may wish to avoid them. Many seed companies now have meadow mixes that include wild herbs.

**Plants for Woodland Edge or Herbal Strips**

| | |
|---|---|
| Agrimony - *Agrimonia eupatoria* | Borage - *Borago officinalis* |
| Black Medick - *Medicago lupilina* | Chamomile - *Matricaria chamomilla* |
| Chickweed - *Stellaria media* * | Chicory - *Cichorium intybus* |
| Coltsfoot - *Tussilago farfara* | Comfrey - *Symphytum officinale* * |
| Common Knapweed - *Centaurea nigra* | Common Sorrel - *Rumex acetosa* |
| Common Vetch - *Vicia sativa* | Cowslip - *Primula officinalis* |
| Dandelion - *Taraxacum officinale* * | Fat Hen - *Chenopodium album* |
| Greater Knapweed - *Centaurea scabiosa* | Hoary Plantain - *Plantago media* |
| Lady's Bedstraw - *Galium verum* | Mallow - *Malva moschata* |
| Marjoram - *Origanum vulgare* | Meadow Vetch - *Lathyrus pratensis* |
| Ox Eye Daisy - *Leucanthemum vulgare* | Parsley - *Petroselinum crispum* |
| Pot marigold - *Calendula officinalis* | Red Campion - *Silene dioica* |
| Ribwort Plantain - *Plantago lanceolata* | Sage - *Salvia officinalis* |
| Salad Burnet - *Sanguisorba minor* | Scurvy Grass - *Cochlearia officinalis* |
| Self Heal - *Prunella vulgaris* | Silverweed - *Potentilla anserina* |
| Stinging Nettle - *Urtica dioica* * | St. John's Wort - *Hypericum perforatum* |
| Sweet Woodruff - *Galium odoratum* | Thyme - *Thymus vulgaris* |
| White Campion - *Silene alba* | Wild Carrot - *Daucus carota* |
| Yarrow - *Achillea millefolium* * | *See also green manure plants (Page 18)* |

** Can be invasive.*

# Companion grazing

All poultry prefer the fresh new tillers of short-growing grasses. These have the best nutritional value. Long grasses, as well as being rank and unpalatable, make hens' feathers wet if the birds are out after rain, and this dampness can in turn be transferred to the nest boxes of laying houses, possibly making the eggs muddy. Coarse grasses can also cause impaction or blockages in the digestive system.

Cutting or cropping the grass produces more tillers, hence the importance of regular topping. The Soil Association standards recommend companion grazing in this context. For example, having sheep on the same range as chickens is beneficial, for the sheep crop the grass, producing new growth suitable for the chickens, while the chickens scratch and disperse the animal droppings, taking advantage of any insects that they disturb in the process. Geese and chickens have traditionally been allowed to graze with cattle, and this was the system followed by my parents and subsequently by me. Some producers prefer to let poultry follow on after the sheep or cattle, rather than letting them share the site at the same time.

Sheep and chickens sharing pasture to their mutual benefit. Some producers prefer to let the sheep graze first, followed by the poultry.

Where companion grazing is not possible, the grass will need to be kept topped, mown or strimmed on a regular basis, to ensure that it is short and with new tillers being produced. If electric fencing or netting is used, the area around it will need special attention to avoid damage from the cutters, although electric netting can be moved temporarily.

Long grass can 'short' the electric current, so it is important to ensure that the area around the fence or netting is not neglected. Some producers lay a plastic strip underneath the fence to keep the grass down, so that the problem is avoided.

## Flock density on pasture

The maximum numbers of birds allowed under UK Organic Standards are:

Layers: 1 hen per 4 square metres. (2,500 per hectare). NB: 1 hectare = 2.47 acres.

Table birds: 1 hen per 4 square metres. (2,500 per hectare) with fixed housing.
1 hen per 2.5 square metres. (4,000 per hectare) with mobile housing.

Ducks: 1 duck per 4.5 square metres. (2,220 per hectare)

Geese - 1 goose per 15 square metres. (670 per hectare)
Turkeys - 1 turkey per 10 square metres. (1,000 per hectare)
Guinea fowl - 1 guinea fowl per 4 square metres. (2,500 per hectare).

The Soil Association Standards specify the following maximum numbers:

Layers: 1,000 per hectare. Table birds: 2,500 per hectare. Ducks: 2,000 per hectare. Geese: 600 per hectare. Turkeys: 800 per hectare. Guinea fowl: 2,500 per hectare.

It should be emphasised that these are maximum numbers. In practice, the numbers need to be far less if, for example, the soil is heavy with a tendency to become boggy, or inclement weather generally has had a negative effect on the land. As referred to earlier, the densities are also far higher than they were traditionally, although there were no standards laid down then, and the flock sizes were far smaller. My parents, for example, never kept more than 50 chickens per acre, and neither did I.

The regulations rather vaguely state that outdoor stocking rates should be 'low enough to prevent poaching of the soil and over-grazing of the vegetation', but given the maximum flock densities they specify, one wonders how one earth this can be prevented! It is not surprising that many organic producers see coccidiosis as the number one health problem! (Ref: *Organic Poultry Production*. Nicolas Lampkin. 1997).

The basic regulations require that all poultry 'should have access to outside ranging whenever weather conditions permit, and for at least one third of their lives'. As referred to on page 9, the Soil Association standards specify that birds have 'continuous and easy access to range during daylight hours'. For layers this should be for all their laying lives, while for other poultry it should be for two-thirds of their lives. Waterfowl are required to have access to a stream, pond or lake. The Soil Association prudently adds that this should be 'well maintained and managed to prevent the build-up of stagnant water and decaying vegetation, pollution and disease risk'.

## Rotation of pasture

Chickens are tough on pasture, and the greater the flock density, the greater the damage. They scratch around in search of insects and invertebrates, and seek out areas in which to make dustbaths, allowing fine soil to trickle through the feathers as an instinctive way of getting rid of parasites. (The Soil Association recommends letting birds choose their own dustbathing areas).

Ducks, with their large webbed feet and frequent, copious droppings, are also quite hard on pasture, while geese consume grass at a rapid rate. Turkeys and guinea fowl are arguably not as tough on grass as the others, but they too leave their marks, and frequent changes of pasture are essential, not only to allow time for the plants to recover, but also to reduce the risk of infection and parasites from over-used land.

Chickens will find their own areas in which to make dustbaths.

The common sense approach is to move a flock to a new area of pasture as soon as it shows wear and tear. This could be as frequent as every week (ADAS recommends every six weeks for free-range birds), but one can see the difficulties involved for some very large producers.

For layers, the Soil Association requires that pasture is rested for a minimum of nine months between batches of poultry, and for table birds at least two months per year, as well as one year in every three years. Where small flocks of up to 50 birds that have a wide roaming area throughout the day are concerned, these requirements do not apply.

How new pasture is made available will depend on the flock size, type of housing and layout of the pasture. The diagrams on page 26 illustrate some of the ways of rotating pasture with moveable and static housing.

Where permanently sited housing is used, the pasture may not always be conveniently situated. In this case, it may be necessary to erect 'races' or temporary fencing in order to direct the poultry to the appropriate area.

## Maintenance of pasture

Reference has already been made to crop systems that include pasture in their rotation, where periodic ploughing and reseeding take place. In the event of permanent pasture, this may not be possible. For example, some pastures may be ancient or protected conservation areas and ploughing would radically alter the environment or habitats of indigenous wildlife.

**Electrifying a Perimeter Fence**

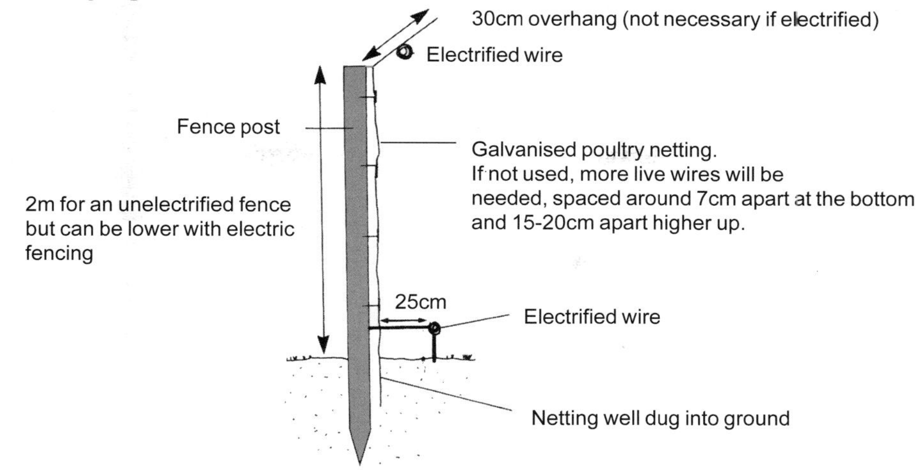


As poultry can be quite damaging of such areas, it is a good idea to check with the local Farming and Wildlife Advisory Group (FWAG) before poultry are allowed access in the first place, in case there are restrictions. Most of the organic standards also prohibit damage to sensitive conservation areas.

Once the poultry have been moved on from pasture, the emphasis is on repair and renovation, with the area resting and recovering. The first necessity is to rake or scarify the grass so that matted growth is loosened and removed. This also has the effect of scattering any areas of droppings and levelling any 'mole-hill' undulations where dust baths have been made.

Poultry droppings are quite acidic and it may be necessary to lime the soil so that the normal pH value of 6.5 is maintained.

Bare areas can be re-seeded and rolled. Once established,they can be accessed by sheep or cattle which will then encourage the growth of new tillers. Reference was made earlier to the fact that leys for poultry pastures are available. (See Pages 19-20).

# Fencing

Organic legislation requires that poultry be protected against predators such as foxes. Shutting the house door at night is a priority, but foxes often appear before dusk when the birds may still be out. Badgers may also be a nuisance in some areas, and roaming dogs, while far fewer in number than they used to be, can still be a problem where there are irresponsible owners. There are only two methods of keeping them all at bay: high perimeter fences or electric fencing.

## Rotation of Pasture

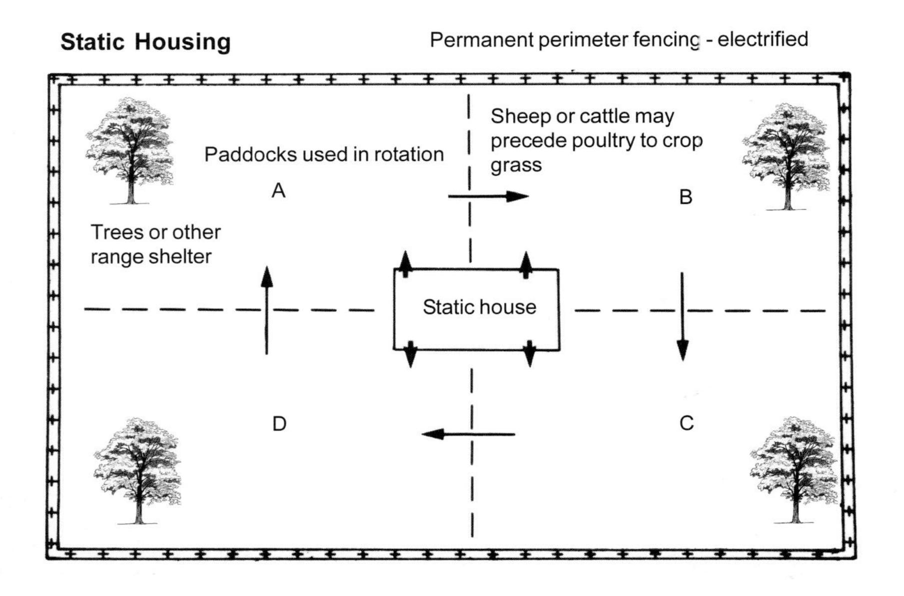


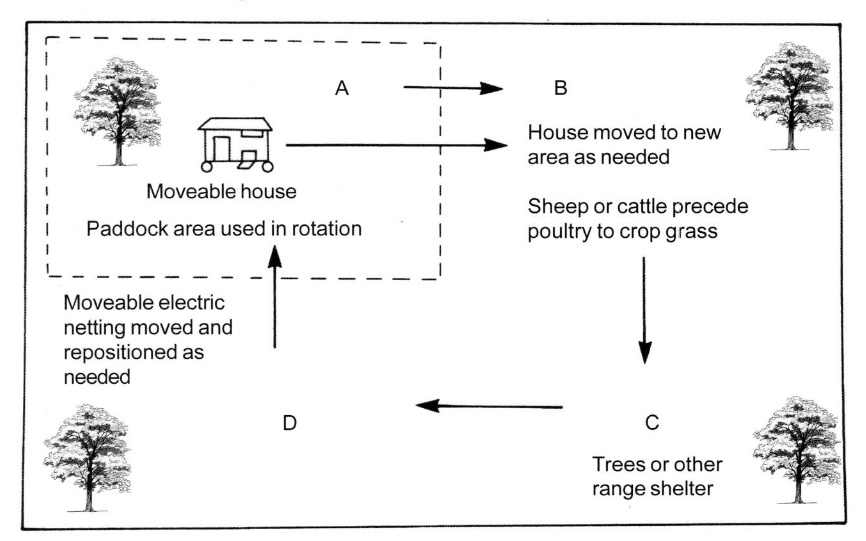


(Ref: *Free-Range Poultry*. Katie Thear. 1990)

Effective fencing is essential in protecting outdoor flocks against foxes. Here an existing fence has been electrified.

## Perimeter fencing

A fence needs to be at least 2m high and ideally have an extra 30cm overhang at an angle facing outwards so that determined foxes cannot scramble over. It also needs to be well dug in to prevent burrowing under. For large areas of pasture, the provision of such fencing might be too expensive, so electric fencing is often used to augment an existing fence.

### Electric fencing

Electric fencing can be used to make an existing hedge or fence secure, or it can be used to make a temporary protective fence. If there is a post-and-rail fence, for example, it can be covered with galvanised poultry netting and then have two electrified wires, at the top and bottom, placed 25cm away from the fence, as shown in the diagram on Page 25.

If galvanised poultry netting is not used to fill the gaps, it will be necessary to have more electrified wires placed horizontally along the fence, spaced around 7cm apart at the bottom and15-20cm apart further up. A similar system can also be used to augment a hedge.

To make a temporary fence, electric poultry netting (*Flexinet*) is ideal, especially where small moveable houses are concerned. It can be erected around the appropriate area and then moved to a new one as needed. The task is much easier if two people are involved. The diagram on Page 26 shows how *Flexinet* can be used with this kind of pasture rotation.

Reference has already been made to the importance of keeping grass growth down under the fence, in case it 'shorts' the current. As weed killer cannot be used in an organic system, the growth has to be restricted by mowing or strimming. A plastic mulch strip under the fence also helps.

## Shelter

Shelter in the form of trees provides shelter from the wind, shade from hot sun, as well as a sense of security for the birds. Chickens are descended from the Jungle Fowl of Asia and are still genetically programmed to be wary of open spaces. Overhead predators such as buzzards or even aircraft represent danger to them. They are more likely to spread out, taking full advantage of the ranging area, if it is sheltered. A well-grassed orchard, for example, is ideal. Organic regulations require shelter provision on the range.

Research by Scottish Agricultural Colleges (SAC), established that in a field without tree cover 55% of a free-range flock stayed in the area around the house, with only 8% of the available pasture being used. Consequently, the area around the house became over-used and the flock density for the area was far higher than it should have been. (Ref: *Bird Performance & Behaviour*. Keeling & Dunn. SAC. 1988). Subsequent research has established that placing shelters on the grass encourages poultry to range more widely. (Ref: *Management Factors Affecting the Use of Pasture by Table Chickens in Extensive Production Systems*. Gordon & Forbes. Proceeding of COR Conference. Aberystwyth. March 1992).

Trees or other forms of shelter some distance from the house have the effect of attracting the birds so that they take full advantage of the available pasture. A distance of 15 to 20 metres from the house is an optimum dis-

Chickens are descended from Jungle Fowl and are drawn to areas of natural shade and protection. *(Premier Poultry)*

tance for such shelter. If trees are to be planted, willows and poplars are quick-growing. Evergreens such as pine and spruce will provide winter shelter, but it is a good idea to find out what trees are appropriate for the local area and soils. It is also important to get right the balance of open area and cover. The aim is to have open pasture with areas of shelter in and around, not completely covering it.

Newly planted trees will need to be protected initially if sheep and cattle are to graze in the area. As they grow, the lower branches can be removed so that a shade canopy is provided, rather than low-lying brush where chickens might be tempted to lay their eggs.

Where there is a lack of natural trees and hedges, it is possible to use thick netting, hurdles, appropriately placed branches or stacked straw bales.

Tall growing game cover crops such as sunflowers, maize and kale are also useful in providing shade and shelter.

A small mobile house on wheels that is easily moved to new pasture. Trees give shade and protection and the hens are protected against foxes by electric poultry netting seen at the back.

With a static house it is more difficult to maintain pasture in a good state for the area immediately outside becomes over-used and muddy, despite the protective weather verandah at the front. *(Onduline)*

# Housing

*What's the use of a house if you can't move it?*
(Welsh organic poultry farmer, 2005)

Housing is of fundamental importance. It provides shelter, security and comfort for the poultry, as well as a place to lay eggs for a laying flock.

There are two options: having a static or fixed house from which the poultry emerge to the same pasture, with areas made available in rotation, or a moveable one that can be moved to fresh pasture. The latter, in my view, is infinitely preferable. A house that is on the same place all the time inevitably has the area around it suffering excessive wear and tear, with mud soon building up outside. This is detrimental not only to the cleanliness of the eggs but also to the health of the birds.

With a fixed house, it is more difficult to organise access to pasture, with flocks needing to be directed to different areas. Often 'races' of temporary fencing need to be erected to control them. This can be a problem if, for example, the Soil Association's requirement that chicken pasture should be no more than 100m (ducks a maximum of 50m) from the house is to be met.

If there is no choice but to use a static house, the ideal situation for it is in the centre of a pasture that can be divided into at least four areas, with access being provided to one area at a time by the judicious use of pop-holes. (See the diagram at the top of Page 26). A verandah or roofed area such as that shown opposite will provide weather protection and prevent mud being taken into the house, although a slatted area, such as that shown on page 33 is also effective.

Providing electricity to a static house is usually easier than it is to a mobile house, although there are now systems available for moveable houses. (See Page 72).

## Housing requirements

### All poultry

UK Organic regulations require that housing for all poultry should be structures with their own air space ventilation, feed and water, and with their own dedicated grazing area. At least a third of the floor area needs to be solid, not slatted, and covered with litter material such as wood shavings, straw, sand or turf. All houses must have pop-holes of a size adequate for the birds, with a combined length of at least 4m per 100 sq.m of the house available to the birds. Maximum area of a house per unit must not exceed 1,500 sq.m.

Inside a large mobile house. The raised area at the back is slatted for the collection of droppings.

## Layers

In houses for laying hens, a sufficiently large part of the floor area must be made available for the collection of droppings. Perch space for them is a minimum of 18cm per bird, with one nest box per 8 birds. Maximum density inside the house for laying hens is six birds per sq.m. Each poultry house must not contain more than 3,000 laying hens.

Natural light can be supplemented by artificial light to a maximum of 16 hours of light per day, with a continuous night rest period of at least eight hours.

## Meat poultry

In fixed housing the maximum density is 10 birds, not exceeding 21kg, per sq.m. In mobile housing which remains open at night, the maximum density is 16 birds, not exceeding 30kg liveweight, per sq.m. Guinea fowl are required to have a minimum perch space of 20cm per bird.

Each poultry house must not contain more than 4,800 table chickens, 5,200 guinea fowl, 4,000 female Muscovy or Pekin ducks (3,200 male Muscovy or Pekin ducks), 2500 geese or 2,500 turkeys.

The above requirements meet the basic housing regulations. The Soil Association, while meeting all criteria, recommends that preference be given to

Outside the same mobile house as that shown on the left. Wide pop-holes are provided for ease of exit and entrance, and the area immediately outside has a foot cleaning area to keep out mud.

mobile houses. They also require that 50% of a laying house floor area be solid, not slatted, by comparison with the basic 30% referred to on Page 31. Pop-holes on both sides of the house are recommended so that adverse weather conditions do not affect the inside of the house. Also recommended is the use of chopped organic straw as litter, replenished regularly and kept in a dry, friable condition. Paper-based bedding is prohibited.

For meat poultry, the maximum stocking density inside is 10 chickens, ducks or guinea fowl, not exceeding 21kg liveweight, per sq.m. For geese and turkeys the maximum density is 2 birds, not exceeding 21kg liveweight, per sq.m, with turkeys needing a minimum perching space of 40cm per birds.

In mobile houses that do not exceed 150 sq.m of floor space and remain open at night, the maximum flock density is 16 table chickens, 16 ducks, 3 geese or 3 turkeys per sq.m.

The Soil Association's recommended maximum numbers of birds per house are: 500 layers, 500 ducks, 500 table birds, 250 geese, 250 turkeys or 500 guinea fowl. However, under certain conditions, they do allow the following maximum numbers: 2,000 layers, 1,000 ducks: 1,000 table birds, 1,000 geese: 1,000 turkeys or 1,000 guinea fowl. As mentioned earlier, it is my view that these numbers are excessive.

## Mobile houses

Moveable houses give greater flexibility to integrate poultry into an organic farming system as a whole. In recent years, there has been a swing towards their use. One of the reasons has undoubtedly been the reluctance of local authorities to give planning permission for large, static houses. There is often a feeling that they are too close in appearance and function to large battery or broiler houses, with all their attendant effects on the environment. Mobile houses, by contrast, tend not to have problems with planners. The need to be able to provide pasture within a recognised rotation system has also made it difficult for producers to justify the use of static houses.

Most organic producers now use mobile houses, but where at one time, these were relatively small, some are built as modules that are big enough to house large flocks. These are mobile only in the sense that they are dismantled and moved once a year, at the end of the season. Some housing manufacturers have in effect produced intensive houses that can be moved!

Far better, in my view, is to use small mobile houses that house a relatively small number of birds, and that are moved several times during a season. These are far more practical and adaptable, ensuring that pasture is used to best effect and that there are potentially less problems such as feather pecking or coccidiosis and other diseases in the flock.

### Design considerations

Most mobile houses can be used for layers or table birds, with the obvious proviso that nest boxes are required for the layers, and perches are needed for the perching birds. The key factors in their design are:

- Sound construction
- Adequate insulation
- Good ventilation
- Appropriate light levels
- Ease of movement
- Appropriate fittings.

Specific considerations, such as the housing requirements for chicks or for layers, are detailed in the appropriate chapters, but the general considerations for all housing are as follows:

#### Construction

Traditional houses are made of wood which is strong, robust and warm. The support timbers are around 3cm thick, with wall cladding that (depending on the manufacturer) is made of one of the following: overlapping weatherboarding, tongue-and-groove boards, or exterior-grade plywood.

A traditional timber house for 100 chickens. It has ridge and side ventilation. There are nest boxes on both sides and pop-holes at each end. It is equipped with skids for moving. *(Gardencraft).*

Inside the same house. Note the hinged perches that can be lifted up during the day. *(Gardencraft)*

The roof may be made of shiplap boarding, wood covered with bitumenised (tarred) felt, or *Onduline* corrugated bitumen. The timber has to be proofed against the weather, ideally by forcing the proofer under pressure.

Traditional wooden houses have stood the test of time, and are often the choice of those with small flocks. Nevertheless, they do have some disadvantages. Cracks in the wood and roofing felt can provide hiding places for red mites which emerge at night to bite and feed on the blood of the birds. Beyond a certain size they can be difficult to move so are not appropriate for those with larger flocks.

In recent years, mobile houses have been designed to be light, easy to move or dismantle, and made of materials that offer good insulation without proving a hiding place for red mites. The chassis is usually made of steel to provide structural strength, with prefabricated and insulated composite panelling. Outer surfaces may be protected with galvanised, plasticised steel.

**Insulation**

The temperature at which chickens, turkeys and guinea fowl operate best is 21°C. Ducks and geese are hardier and have more insulating down feathers in their plumage. All birds provide the warmth themselves, while the house retains it by having adequate insulation. Just as important is the need to avoid over-heating which can take place in hot summers. Having reflective insulation in the roof and walls helps to keep the environment comfortable, while a maximum/minimum thermometer in the house enables regular checks to be made. Reference has already been made to the importance of insulating materials not providing a refuge for red mite. Equally important is the need to balance insulation with ventilation.

**Ventilation**

A combination of ridge and side ventilation is ideal. Cold air enters through inlets in the walls, rises as it becomes warmer, and exits through the roof ridge outlet. Wall inlets may be windows or vents placed in such a way that perching birds are not subjected to draughts. The vents can usually be opened enough to cater for specific birds, with waterfowl generally needing a greater rate of air flow, as referred to earlier. Some houses are equipped with automatically operated vents, rather like those that are often used in greenhouses.

**Light levels**

Natural light coming into the house is necessary, but it needs to be diffused or reflected. Nest boxes, for example, need to be in shady areas of the house to make them more attractive as laying areas, while too much direct light can also foster aggressive behaviour or egg-eating.

One of the range of mobile houses designed to be easily dismantled for ease of cleaning and moving. *(Associated Poultry)*

The house in the process of being dismantled. The smooth interior ensures that there are no hiding places for red mites. *(Associated Poultry)*.

When natural daylight decreases in winter, it is necessary to supplement it with artificial light if layers are to continue laying. This should not exceed a total of 16 hours a day, with a minimum of 8 hours darkness being available at night. Artificial light can be provided by mains electricity or by means of 12 volt systems. (See Page 72).

### Ease of movement

Generally speaking, traditional timber houses for around 100 birds are the largest that can be moved on skids, with a tractor or 4x4 vehicle with towbar being needed to drag it.

Where wheels are used, they need to be sufficiently wide and strong to cope with the weight. Many mobile houses are now designed to be dismantled so that both cleaning and moving are facilitated.

### Appropriate fittings

Layers will need nest boxes and the best type are those that are accessible from outside the house. Those that are equipped with 'rollaway' facilities ensure that the eggs roll towards the collecting area where they cannot be reached by the hens. Some houses have the facility for closing off nest boxes at night so that the hens do not sleep in them.

Perches are best placed at a higher level than nest boxes, but below the side vents so that birds are not in a draught. They need to be easy to remove for ease of cleaning.

Some houses have hinged perches that can be put up during the day, to maximise house space and provide ease of access to nest boxes. They also remove the temptation of providing perching space during the day when the hens should be concentrating on laying if they are in the house.

Many of the new mobile houses have a wide range of optional extras, including being fitted with automatically filled feeders and drinkers as well as 12-volt lighting systems.

## Different types of houses

Reference has already been made to the fact that most houses, with a little adaptation, can be used for different types of birds. Chicks may be brooded in the same house that they subsequently have when grown, or they may be in a separate house from which they are moved. If this is the case, it is important to ensure that temperature and ventilation are equivalent in both houses, and stress during transference is avoided.

A house made of galvanised metal with plasticised insulation on the inside. Ventilation and light are provided by the vent and the wide pop-holes. A drinker is provided outside as well as inside.

Inside the same house. The suspended feeders and drinkers are filled automatically.

Feeding grain on pasture encourages the flock to range over all the area provided.

Open feeders need to be heavy-based to avoid being turned over.

# Feeding

*It is impossible to attach too much importance to foods and feeding.*
(Herbert Howes. *Management of Farm Poultry*. 1930)

All living creatures need regular inputs of food in order to fuel the body metabolism. Food consists of proteins, fats, carbohydrates, water, minerals and vitamins, and these enable and maintain growth, energy and health.

Poultry, including chickens, ducks, geese, turkeys and guinea fowl, are naturally omnivorous, eating grains, seeds, invertebrates, small vertebrates and plants. Free-ranging birds will find a proportion of these on the pasture and in hedgerows, but the basic requirements are provided by giving them a balanced ration that is suitable for their particular needs. This may be by feeding an all-in proprietary organic feed or by giving the birds a range of different feed materials that have not been compounded (straights).

Organic standards specify that poultry must be fed on organically produced feedstuffs containing roughage, although a certain proportion may include in-conversion and non-organic feed ingredients. This reflects the reality of the situation where there is sometimes a shortage of organically produced foodstuffs. At the time of writing, non-organic ingredients are due to be phased out by the beginning of 2012. It is important to bear in mind that standards are subject to amendment.

The standards specify the list of acceptable food materials, including those of plant origin such as cereals and pulses, those of animal origin, such as milk, fishmeal and eggs, and acceptable sources of minerals and vitamins. They are of prime concern to those who are producing compound rations. Most organic producers use proprietary rations. Only those with sufficient land are in a position to grow all their own feed straights.

No genetically modified ingredients are permitted in organic rations. Also banned are animal by-products and in-feed additives such as antibiotics, coccidiostats and growth promoters.

Balanced rations are based on the following, with ingredients and quantities varying according to the type of feed required: wheat, barley, maize, oats, peas, beans, soya, triticale, linseed, lucerne, minerals and vitamins. Compound feeds are available as starter, grower, finisher and layer rations, and are formulated accordingly. Organic wheat and mixed grain, which is wheat with a proportion of maize, are also available from suppliers.

# Feed requirements

Different feed components cater for differing metabolic requirements. They are normally grouped as proteins, energy feeds, minerals and vitamins.

## Proteins

These are the main body builders, responsible for growth and maintenance of body tissues. Proteins are made up of various constituents called amino acids. There are around a dozen of them but the most important are lysine, methionine and tryptophan. These three have to be taken in directly every day, while the others can be synthesised from existing food constituents.

Amino acids are found in animal sources and to a lesser degree in plants. Non-organic compound feeds usually include synthetic amino acids, but these are not used in organic feeds. Fishmeal is an acceptable organic source, where this can be supplied by a registered mill, but there have been some consumer complaints that its inclusion produces a fishy taste in eggs. Soya contains lysine, methionine and tryptophan.

Free-ranging birds will cater for some of their needs by eating insects and other invertebrates, and occasionally small vertebrates. Researchers in Sweden have suggested that encouraging earthworms in mulch strips could provide a food source. (Ref: Ekstrand & Elwinger. 1996). Plant sources include protein from potato skins, maize gluten (sometimes called prairie meal), boiled linseed expeller, and soya expeller. Expeller is the by-product of oil extraction during the crushing of plant material. Where extraction is used, there is a requirement for this to be by physical extraction rather than by using chemical solvents. Dried yeast and milk products are also sources of amino acids.

The need for proteins, particularly lysine, is at its highest in chicks, with the need gradually reducing with age. Starter rations contain around 19-22% protein (depending on the birds), while grower and finisher rations are typically around 15-16% protein. Layer rations are usually around 16%.

**Protein Sources**

| | |
|---|---|
| Animal | invertebrates and small vertebrates from pasture ranging, fishmeal, milk products |
| Plant | soya, peas, beans, barley, maize, sunflowers, barley, oats, wheat, clover, grass, lucerne, dried yeast, potatoes, linseed |

## Energy feeds

As the name implies, these are the feed constituents that supply most of the energy requirements associated with everything from moving about, laying eggs and anything else associated with being a living, breathing creature.

Carbohydrates and fats (oils) are the main suppliers, with cereals and soya being the main sources. Energy in food constituents is called the metabolisable energy (ME) and is measured in terms of megajoules per kilogram (MJ/kg). The normal laying hen, for example, would require 11.5MJ/kg in her daily ration.

**Energy Sources**

Wheat, barley, triticale, maize, oats, millet,
soya, sunflowers, linseed, molasses

## Minerals

Minerals are inorganic materials (meaning that they are not sourced from living animals or plants, not that they are unacceptable to organic standards). A wide range is required, some in minute quantities, hence referred to as trace elements. Some are needed in larger quantities. Calcium and phosphorus in the form of limestone are needed for bone formation and egg shell production. Limestone is normally added to feeds. Calcified seaweed made available separately is also a source. Birds will not take more than they need. All minerals play their part in maintenance of health and guarding against the onset of nutritional deficiency diseases. Free-ranging birds will acquire some minerals from their foraging activities, but minerals are normally added to compound feeds.

**Natural Mineral Sources**

| | |
|---|---|
| Calcium | lucerne, green vegetables, calcified seaweed, milk products, molasses, dried yeast, soil (if in local soil) |
| Cobalt | calcified seaweed |
| Copper | cereals, beans, soil (if in local soil) |
| Iodine | calcified seaweed, washed seaweed |
| Iron | nettles, chicory, parsley, dandelions, chickweed |
| Magnesium | oats, beans, soya, grass, spinach |
| Manganese | lucerne, wheat, maize, millet, oats, molasses, dried yeast |
| Phosphorus | lucerne, grass, oats, dandelions, calcified seaweed |
| Potassium | lucerne, maize, wheat, soya, sunflowers, potato, dried yeast, molasses |
| Selenium | green vegetables, lucerne, maize, dried yeast |
| Sodium (chloride) | grass, lucerne, maize, molasses, dried yeast, sunflowers |
| Zinc | lucerne, soya, molasses, wheat, maize, sunflowers, dried yeast |

## Vitamins

Vitamins are organic compounds, as distinct from the inorganic minerals. They are essential for the maintenance of health and many of them will be sourced through the free-ranging activities of the birds. They are also added to compound feeds.

**Natural Vitamin Sources**

| | |
|---|---|
| Vitamin A | grass, maize, kale & brassicas, carrots, nettles |
| Vitamin B1 (Thiamin) | most cereals |
| Vitamin B2 (Riboflavin) | grass, soya, dried yeast |
| Vitamin B3 (Niacin) | wheat and other cereals |
| Vitamin B5 (Pantothenic acid) | grass, comfrey, dried yeast, molasses |
| Vitamin B6 (Pyridoxine) | soya, wheat, dried yeast |
| Vitamin B12 (Cobalmin) | comfrey, calcified seaweed |
| Vitamin D3 | sunshine, wheat, maize and other cereals |
| Vitamin E | grass, wheat, maize, kale and other brassicas |
| Vitamin H (Biotin) | maize, lucerne, grass |
| Vitamin K | lucerne, grass |

## Grit

All poultry must have grit in order to break down grains within the gizzard, a specially adapted region of the digestive tract.

Insoluble grit is available from suppliers and can be made available in a container from which the birds help themselves as required. Free-ranging birds will also find their own sources of small stones.

## Compound feeds

Compound or proprietary feeds are formulated to contain all the necessary nutrients for specific birds and situations, although it is quite common for a producer to use the same organic ration for a range of different birds. For example, chickens, ducks and geese are usually fine on chicken starter and finisher rations. Turkeys and guinea fowl need more protein initially so a specific turkey starter ration is best for them, but they can be given a chicken finisher ration afterwards.

Mills that are producing organic feeds must be registered to do so and are required to use only those ingredients that are acceptable to the standards. Depending on what ingredients are available at any particular time, the composition of the feeds may vary, and the relative cost may reflect this.

Buying in bulk is obviously cheaper, but adequate storage facilities will be required and it is important to remember that compound feeds do become stale over a period of time. Buy only what you know will be used before this happens!

Typical ingredients used for organic compound feeds include: organic wheat, barley, wheatfeed, peas, alfalfa pellets, non-GM soya, non-solvent extracted soya oil, molasses, lucerne, linseed, sea salt, calcined magnesite, calcium carbonate, dicalcium phosphate, yeast, wheat germ, natural vitamins. (Alfalfa pellets are lucerne).

A grit feeder from which poultry can help themselves as needed. This one is inside a static house for ducks. Note the Yorkshire boarding that provides plenty of ventilation above head level.

As referred to earlier, compound feeds are available as:

- *Starter* rations for chicks
- *Grower* rations to follow on
- *Finisher* rations for the final period of table bird production.
- *Layer* rations for laying birds
- *Breeder* rations for parent birds (with an increased level of minerals to prevent deficiency diseases in the chicks)

They are produced as *pellets, mash* or *crumbs*. Pellets are easy to feed and store but are more expensive than mash which is the powder form. The latter is more appropriate for some automatic feed systems, although 'flicking' by the birds can represent more wastage. The calcium content may also settle at the bottom so that it is temporarily unavailable to the birds until stirred.
(Ref: *Calcium Distribution and Environmental Studies*. Norman & Magruder. Cargill Inc. 1965)

Compound feeds normally have the maximum amounts of the following ingredients shown as percentages:

- *Oil* for growth and energy
- *Protein* for growth and body maintenance
- *Fibre,* the cellulose or roughage content
- *Ash,* the mineral content

Wide vents and pop-holes provide light and access, while suspended feeders and drinkers are available for the chickens. *(Hubbard-ISA)*

**Examples of Organic Compound Feeds**

**Chicken**

| *Starter* | *Grower* | *Finisher* | *Layer* |
|---|---|---|---|
| Oil: 5.4% | Oil: 3.10% | Oil: 4.2% | Oil: 4.0% |
| Protein: 18.0% | Protein: 13.8% | Protein: 16.0% | Protein: 17% |
| Fibre: 4.3% | Fibre: 3.9% | Fibre: 3.7% | Fibre: 4.4% |
| Ash: 5.8% | Ash: 5.6% | Ash: 5.1% | Ash: 13.0% |

*(Reference: Marriages Feeds)*

## Grain

Feeding a whole grain ration is not strictly necessary if a compound feed is being provided, but there are good reasons for doing so. Poultry like it and it gives chickens, guinea fowl and turkeys a chance to scratch about to retrieve it if it is placed directly on the ground. In this way, one of their innate needs to scratch is being met. Another important reason for feeding grain in this way is that it helps to avoid or reduce the incidence of problems such as bullying within the flock.

Ducks and geese are also partial to grain and it is a useful incentive to get them into their house at dusk, if they are reluctant to go inside.

Many producers prefer to make grain available in a feeder. Placing this in a different area each time helps to ensure that all the available pasture is being used. Outside feeders can also be used for compound feeds.

An outside feeder that is moveable and protected against the weather. The turkeys can help themselves as required with no danger of wild birds stealing the food. It can be lowered for use by other birds. *(Hengrave Feeders)*

In winter, when more demands are placed on the system, feeding extra grain is a cheaper alternative to increasing the compound feed ration. Laying hens, for example, need an extra 4.2 calories for every $1^{\circ}$ fall in temperature from the optimum of $21^{\circ}$C. (Ref: *Feed Intake of Free-range Birds in Relation to Temperature.* Warner. ADAS. 1987). Increasing the amount of compound feed at this time is to be avoided, not only on economic grounds, but also because the excess protein is merely expelled in the faeces.

Wheat is the best choice for all the birds. Giving mixed grain is more expensive. It is predominantly wheat with a proportion of maize but greedy birds tend to pick out the maize, and may become over-fat as a result.

It is important to bear in mind that too much grain feeding is to be avoided if a compound ration is being given, otherwise the overall balance of nutrients is affected. Typically, 15g per bird, per day, is sufficient for laying hens, with an increase to 20g per bird, per day in winter. As referred to earlier, insoluble grit must be made available.

## Farm-produced feeds

Most producers will be buying in organic compound rations, but those with larger areas of land may wish to grow a proportion of their own feed crops, especially if these are part of an overall crop rotation system. All the crops will need to be processed, either mechanically in order to separate the grain or to heat-treat the legumes. Adequate working and storage space is obviously needed. It should be emphasised that home-mixing of poultry feeds is a skilled operation and not one to be undertaken without advice. Adequate milling and mixing facilities are required, as well as a detailed knowledge of achieving the right balance of nutrients, bearing in mind that some added ingredients are banned under the organic standards.

Some producers grow feed crops organically but as they lack milling and mixing facilities, they supply their crops to organically-registered mills which then process them for use in compound feeds. The feeds are then supplied back to the producer.

### Wheat

This is the main energy feed in poultry rations and is used rolled, ground or crushed in compound feeds. Whole wheat can be given as a grain feed on pasture, as referred to above.

### Barley

Barley is high in energy but relatively low in vitamins. It is used in poultry feeds where it is balanced with other feedstuffs. It can be rolled, crushed, ground or flaked. It frequently figures in crop rotations.

### Oats

Oats are very high in fibre so are not widely used in poultry feeds. However, recent breeding developments have produced hull-less naked oats which are lower in fibre and higher in oil and other nutrients. It is now being grown as a constituent of poultry feeds. (Ref: *Avian Feed Efficiency from Naked Oats*. Valentine & Cowan. IGER. 2004).

### Triticale

Triticale is a good energy-providing cereal that also has lysine and methionine amino acids. It can be grown to provide a grain feed to accompany a compound ration, or as a constituent of a mixed feed.

### Maize

Known as corn in the USA, most of the maize in poultry rations is bought in because, apart from in a few sheltered areas, it does not ripen properly in the UK. Maize gluten, usually called 'prairie meal', forms part of the non-organic constituents of compound feeds and is a source of protein and energy. It also adds a golden colour to the carcases of 'corn-fed' table chickens.

### Lucerne

Lucerne (alfalfa in the USA) is a tap-rooted herbaceous perennial that lasts for around 5 years. High in protein, minerals and vitamins, it is extensively used in the USA for pastured poultry. It also enhances the colour of egg yolks. Where it is grown in the UK, it tends to be in the warmer, drier areas of the

Automatically replenished feeders with manually filled drinkers in a small mobile house being used for chick rearing,

south-east. It is processed for use as lucerne meal, a compound feed constituent, or to produce alfalfa cubes. In milder areas there is no reason why it should not be grown as poultry pasture, as part of a rotation.

### Peas

Like all legumes, peas add nitrogen to the soil and are a useful part of a crop rotation. They are used to provide protein and energy constituents in compound feeds, but need to be heat processed to ensure full digestibility.

### Field beans

This legume also adds fertility to the soil within a crop rotation. It provides protein and energy constituents in compound feeds but needs to be mill-processed to maximise its potential.

### Lupins

Agricultural lupins are good sources of protein but are relatively low in the amino acids that are necessary in poultry diets. They need to be processed to remove the hard shell cases.

### Soya

Most soya is imported and there is an organic requirement that it should not be genetically modified. Full fat soya is the full bean where the oil has not been extracted. Soya expeller has had the oil removed, but to meet organic standards it needs to have been pressed rather than to have been subjected to solvent extraction.

(Continued on Page 52)

## Laying Hens

Bovans Nera

White Star

Hebden Black

Speckledy

## Table Chickens

Cotswold Golds

Sasso

Sherwood White

Hubbard-ISA

### Fishmeal

Fishmeal is used by some compounders and is accepted by organic standards as long as it comes from non-intensively run fish farms. It is used in small quantities only and is a source of protein and minerals. Many producers prefer not to use it because of consumer complaints that a fishy taste is imparted to the eggs of layers that are fed on it.

### Potatoes

Vegetable protein from potato skins is sometimes used by compounders but it is not widely available from organic sources.

### Yeast

Traditionally, dried yeast has always been valued by poultry keepers as a good source of amino acids, vitamins and minerals. It is often included in compound rations.

### Molasses

Molasses provide energy and small amounts of vitamins and minerals, as well as having a binding effect on powdery ingredients in a compound feed.

### Vegetable oils

Small amounts of expelled vegetable oils may be added to compound feeds, although the use of full fat soya will decrease the amounts required.

### Milk products

Milk and milk products such as dried milk are useful sources of protein, vitamins and minerals, but too much lactose can cause digestive upsets. Traditionally, small amounts of skimmed milk after butter churning were used.

## Feeders and drinkers

It is important to site feeders and drinkers in such a way that spillages are avoided so that damp, soiled areas do not provide a haven for disease-causing organisms in the litter. Siting them over a slatted area helps to avoid this. Moving outside ones is important for the same reason as well as to encourage wider ranging over the whole pasture area.

Containers are available in galvanised metal or heavy duty plastic, and can be used on the ground or suspended. Lightweight plastic ones are more appropriate for inside use. Hoppers or tube feeders are popular with many producers, while bell drinkers are commonly used. These are suspended in the house and are automatically filled from a header tank. Manually filled drinkers need to be replenished daily.

Purpose-made feeders are available for outside use. These have lidded weather protection and some are designed as 'self-feeders' so that, by displacing a lever, the birds can release a small amount of feed at a time. This not only avoids wastage but also prevents food being taken by wild birds. It is important that feeders and drinkers are cleaned regularly and if placed outside they should be in shaded areas.

## Feed conversion ratio

The feed conversion ratio is the amount of feed given in relation to the body weight gain of table birds or the egg mass produced by layers. It is important to know what the feed conversion ratio is in each case, otherwise the profitability of an enterprise cannot be judged.

In order to work it out, it is necessary to keep careful records of the amounts of feed purchased and used, and to take sample weighings from a batch of table birds. It is not necessary to weigh them all, but a few to indicate the average weight gain for the batch. In the case of layers, egg numbers and grade weights should be recorded.

The following gives an indication of the amounts of food consumed and the relative performance of the various poultry, but it should be regarded as a generalisation only, for much depends on type, strain, sex and environmental conditions.

**Compound Feed Consumption per 100 Birds**

| | *Starter* | *Grower* | *Finisher* | *Layer* |
|---|---|---|---|---|
| *Chicken* | 150kg | 300kg | 400kg | 14kg/day |
| *Duck* | 150kg | 300kg | 400kg | |
| *Goose* | 165kg | 750kg | 1200kg | |
| *Turkey* | 175kg | 800kg | 1500kg | |
| *Guinea fowl* | 130kg | 250kg | 350kg | |

**Bird Performance under Free-Range Conditions**

| | |
|---|---|
| *Laying Chickens* | Eggs: 300 - 330. Flock life: 1-3 years. |
| *Table Chickens* | Liveweight at 84 days 2.5kg – 3.5kg. |
| *Laying Ducks* | Eggs: 280 - 330. Flock life: 3 - 4 years. |
| *Table Ducks* | Liveweight at 84 days: 3.0 - 4.0kg. |
| *Geese* | Liveweight at 98 days: 4.0 - 8.0kg. |
| *Turkeys* | Liveweight at 168 days: 6.0 - 15.0 kg. |
| *Guinea Fowl* | Liveweight at 94 days : 1.0 - 1.5kg. |

Commercial Pekin strain of table ducks.

Legarth strain of Embden geese.

Bronze turkey stag displaying. This is a pure breed used for breeding purposes.

Lavender and Pearl Guinea fowl.

The only eggs that are covered by the Egg Marketing Regulations are those laid by chickens.

# Layers

*As sure as eggs is eggs!*
(Traditional saying)

Despite the quotation above, eggs are not necessarily eggs, it seems. The Egg Marketing Regulations, which all producers must meet whether they are organic or not, only apply to the eggs of chickens. The eggs of other birds must obviously be produced and sold in clean conditions, for they are food products, but they are not covered by the other regulations.

Most eggs that are sold are chicken eggs, but there is also a small market for duck eggs and for those of geese, especially for egg-decorating.

## Housing

Small flocks in small mobile houses that have extensive ranging areas are far less likely to have problems than those that rigidly adhere to the maximum allowed under the standards. These allow a maximum of 3,000 birds per house, with one square metre being allowed for every six birds, and at least a third of the floor must be solid not slatted. The Soil Association standards specify 500 layers per house with the same stock density, but with at least half of the floor area being solid not slatted. Under certain conditions, they do allow a maximum of 2,000 birds in a house but that derogation must be applied for. In my view, a maximum of 50 birds per house is the ideal!

For laying hens there are obvious adaptations that need to be made to a house, not only to cater for their needs, but also to make provision for their role as egg layers.

### Floor litter and scratching area

The solid floor area of the house needs litter, not only to absorb any droppings (although most of these are deposited under the perches), but also to provide a scratching area while birds are inside. This is a requirement of all the organic standards, in the event that bad weather conditions make hens reluctant to go outside. Wood shavings from untreated sources, straw or sharp sand can be used. Although unchopped straw is allowed, it tends to mat and is difficult to keep in a friable, aerated condition. Such litter becomes damp, providing a breeding ground for pathogens, particularly coccidiosis organisms. Ammonia gas leading to respiratory problems is also produced. Floor litter in the scratching area should be raked regularly to get rid of damp areas and also to introduce air. Soiled litter should be replaced regularly.

(Continued on Page 60)

Large pop-holes ensure that these organic layers have easy access to pasture from their mobile house.

Insulated mobile houses for young table chickens. They exit through a door rather than a pop-hole.

A mobile house with trees for shade and electric poultry netting for protection. *(Associated Poultry)*

A small mobile house that has had extra rollaway nest boxes placed on the left side.

Here the nest boxes are partially covered with hanging curtains to provide a darkened area that helps to prevent egg eating. The automatically filled drinkers are placed above a slatted area. *(Patchett)*

In very small houses, a sheet of plastic under the litter makes regular cleaning easier, for it is then merely a matter of taking out the sheet with droppings and emptying it onto the compost. Alternatively, there may be a droppings board that slides out from under the perch.

Slatted areas are designed to allow droppings to fall through to a collection area underneath. This may be to a droppings box or pit, or to the ground underneath, for clearance when the house is moved.

## Perches

Perches need to be 4-5cm wide and with slightly rounded corners. These provide the best 'gripping' for perching birds whose feet are equipped with a kind of 'automatic lock' so that they do not fall off when asleep. 18cm is the minimum space required for each bird on the perch, with perches spaced at least 45cm apart and no higher than 60cm from the nearest take-off surface. Where nest boxes are off the ground, it is appropriate to have alighting perches in front of them.

Perches should be easily removed for cleaning, for red mite can find a home in wood cracks. Alternatively, perches made of synthetic materials are available. Some producers use so-called integrated perches which are in ef-

The perches on the left are above droppings boards that can be pulled out for cleaning. The hinged flight rails on the right are to provide ease of access to the nest boxes. At night they can be raised to prevent hens sleeping in the nest boxes. *(Smith Sectional Buildings).*

fect, just part of a slatted floor. They are accepted by the organic standards, but I dislike them because they are just thickened rails slightly raised from the floor. Hens instinctively go upwards to perch and will often try to reach a higher structure, such as nest boxes, if they can. (Ref: *Animal Welfare*. Olsson & Keeling. 2002). If they cannot, the result is likely to be increased levels of stress.

## Nest boxes

The organic regulations require there to be at least one nest box for every eight birds. (The Soil Association standards specify one nest box for every six birds). In small houses, my experience is that one nest box for every three birds is best, with a dimension of 30cm x 30cm being appropriate. If there is a queue, eggs are likely to be laid on the floor.

We all know that hens like to lay eggs in a dark, sheltered area, and that they also show a preference for a lining material that they can shape into a nest. Indeed research has borne this out. (Ref: Bauer, Fölsch & Hörning. 1994). Nest boxes should therefore be in a shaded area of the house, with alighting perches for ease of access if they are above the floor. Having light shining directly into the nest boxes can lead to eggs being laid on the floor in a darker area, or provoke incidences of egg eating.

(Continued on Page 64)

A laying flock of Hisex Rangers out on pasture.

Automatically filled outside drinkers here being used by Cotswold Gold table chickens.

Shade and wind protection, as well as running water, are available for these laying ducks.

Commercial Kelly Bronze turkeys on range with straw bales providing perches as well as shelter and wind protection. *(KellyTurkey Farms)*

Rollaway nest boxes keep the eggs clean and prevent egg-eating but do not cater for the hen's preference for a natural lining material.

Every effort needs to be made to keep the eggs as clean as possible, so nest box liners should be checked regularly and replaced if necessary. Wood shavings or chopped straw are commonly used. Having a slatted area, whether inside the house or just outside the pop-holes, helps to keep mud out of the house and hence away from the eggs.

Rollaway nest-boxes are useful in that, once laid, the egg rolls away to a collecting area that the hen cannot reach so egg eating is also prevented. A synthetic material is often used as a liner so that it does not impede the ability of the egg to roll away, but this does not equate with the hen's preference for a natural nesting material, as referred to earlier.

In small houses, nest boxes that can be accessed from the outside are convenient for egg collection.

## Pop-holes

Pop-holes for entry and exit of hens, to and from the house, are required by all the standards to provide a minimum length of 4m for every 100sq. m of floor area. A convenient size for a medium-sized mobile house, for example, would be 45cm high x 100cm length. This ensures that there is plenty of room for several hens to go through at a time. Small traditional houses would not normally have pop-holes as wide as this, but as their floor area is much smaller, there are no potential problems about going in and out.

The Soil Association recommends having pop-holes on both sides of a house so that access to pasture is more easily controlled, and also prevailing winds can be kept out of the house in the event of poor weather.

Nest boxes that are accessible from the outside are convenient for egg collection. In this house, the side wall opens as a sliding door.

## Breeds

Traditionally, poultry farms bred their own replacement layers, often using pure-bred birds to produce crosses. The Rhode Island Red male crossed with a female Light Sussex, for example, produced sex-linked chicks that could be identified as male or female at hatch. The silvery white males were raised for the table, while the golden brown females became the new laying hens. (Incidentally, the cross is not sex-linked the other way round).

In the past, the most productive layers were the Rhode Island Red for brown eggs and the White Leghorn for white eggs. These two breeds have been instrumental in developing most of the hybrids in today's egg sector. Examples are the Warren-ISA for brown eggs and White Star for white eggs.

Another popular cross was the Rhode Island Red male crossed with female Barred Plymouth Rocks. The chicks are all black but can be differentiated by the fact that the males have head spots. This particular cross, often known as the Black Sex Link is used in the production of Black Rocks and Hebden Blacks in the UK and Bovans Nera in the Netherlands.

It is all too easy to believe that just by crossing two pure breeds such as the ones referred to above, it is possible to breed good layers. Unfortunately, it is not the case. The breeding birds must be from good utility strains with a good record of egg production in their lines.

Rhode Island Red male, a breed of major influence in the development of egg breeds.

Barred Plymouth Rock hen, traditionally one of the best layers and producers of laying crosses.

Most pure breeds today have been bred for show purposes rather than for production. They may look beautiful but they do not lay the number of eggs that would be acceptable commercially. The old saying that 'strain is more important than breed' holds true. It is not all doom and gloom, however, for there are breeders who are concentrating on the utility aspects of the old breeds with the aim of improving their production. The Utility Poultry Breeders' Association is currently working to this end.

The value of pure breeds has also been recognised in recent years by those who are using them to cross with more productive strains to produce layers suited to free-range conditions, or for the production of speciality eggs.

The ideal situation for the organic egg producer is to produce his own replacement stock, for the birds are less subject to stress and are able to acquire and build up immunity to disease for a particular site. However, it is not realistic for many people so the birds must be brought in. Breeding also requires the acquisition of first-rate breeding stock and this can be difficult to find. No birds can be bought from caged systems.

Organic standards require that birds are organically reared but, if they are not available, non-organic birds can be brought in, ideally before they are three days old. By derogation, laying pullets can be introduced later as long as they are not older than 18 weeks, and a conversion period of 6 weeks is allowed before eggs are sold as organic. Ideally a younger age is preferable, to give them time to settle down before starting to lay.

As to the choice of breed, this is a personal one and may well be dictated by the proximity of a suitable supplier.

The blue-green egg shells of Araucanas are popular as specialities.

Light Sussex. One of the traditional dual-purpose breeds but a good utility strain is needed.

## Choosing a breed

The factors to bear in mind when choosing a laying breed are as follows:

- They should come from an organic rearer.
- They should be the same age and size.
- They must not be beak-trimmed.
- They should be used to perching and have the ability to forage.
- They should be healthy and free of external parasites.
- Production data should indicate good potential performance.
- They have been bred for increased docility.
- No wing clipping is allowed.

Most suppliers will provide production data and management details, and should be prepared to replace any fatalities that occur within a short time of arrival. The aspect that birds should ideally have been bred for increased docility is important, bearing in mind that beak trimming is not generally allowed as a means of controlling aggression within the flock. Clipping the flight feathers on one wing to control flightiness is also barred. Birds that are the same age and size are more likely to integrate as a flock.

## Hybrids bred for free-range

In recent years, breeds have been bred specifically for free-range conditions so they are slightly heavier than those destined for indoor conditions. Some have been bred in Britain while others have been introduced from other parts of Europe, including France, Netherlands, Germany and Belgium. They are hybrid crosses that are based on the following traditional laying breeds.

## Traditional Laying Breeds

| | | |
|---|---|---|
| Rhode Island Red | Barred Plymouth Rock | Light Sussex |
| White Sussex | White Leghorn | Black Leghorn |
| Buff Leghorn | White Wyandotte | New Hampshire Red |
| Marans | Welsummer | Cream Legbar |

The type of crossing determines the plumage colour, which may be predominantly brown, black, white or grey (blue), as well as the egg shell colour which includes white, various shades of brown, and blue-green. The level of production will also have been determined by selective breeding of particularly productive strains.

Breeders obviously give their own names to their birds, so that for example, they all have a white egg layer based on the White Leghorn, with names such as Bovans White, ISA White, or Hy-Line White. Suppliers who may be selling a range of chickens from different sources may also give their own names to the birds they are selling, so the same bird may be available from several sources under different names. Pullets may or may not be available as organically reared, depending on the supplier, although in some cases, it is possible to have them to order.

### Black feathered

Most of the black feathered layers are the black sex-link birds referred to earlier, and are produced from Rhode Island Red x Barred Plymouth Rock.

*Black Rock* This has dense, tight feathering in order to cope with outdoor conditions. It produces 280+ brown eggs.

*Bovans Nera* This Dutch breed is lighter in weight and lays 300+ brown eggs.

*Hebden Black* Although it is also a black sex-link, there is a proportion of lighter feathers, particularly at the front of the body. This docile bird's production is around 280-290 dark brown eggs.

### Brown feathered

Most of the brown feathered birds have Rhode Island Red and White Leghorn in their ancestry, and sometimes strains of other breeds such as Light Sussex or New Hampshire Red, depending on the strains.

*ISA-Warren* Based on the original Warren hybrid, this is bred for free-range and is a prolific layer of 300-330 brown eggs.

*Bovans Goldline* This Dutch strain can lay 300-330 brown eggs.

*Hisex Ranger* Production is around 300 brown eggs.

*Lohmann Tradition* Developed in Germany, this lays 300+ light brown eggs.

*Calder Ranger* Bred for increased docility, it lays 300-310 brown eggs.

*Babcock 380.* Available under several names, depending on the seller, its production level is 300+ brown eggs.

*Rhode Star* A hybrid or commercial strain of Rhode Island Red, it lays around 260 brown eggs.

*Maran Cuivree* Also called Starlight or Copper Maran, this is a French import suitable for speciality production. It lays around 200 chocolate coloured eggs.

**White feathered**

Most of the light hybrids are based on the White Leghorn while heavier ones are based on the Light Sussex or the White Wyandotte.

*White Star* Orginating in the Netherlands, this lays 330+ white-shelled eggs. It can be quite flighty so needs high fences.

*Amber Star* This is said to be more docile. The plumage has amber tinges and the breed lays around 300 light brown eggs.

*Sussex Star* Based on the Light Sussex with some Cobb input, this comes from Belgium and is regarded as a dual-purpose bird, laying around 240 cream eggs, but it is also suitable for the table.

**Grey feathered**

*Speckledy* Based on the traditional Marans, the Speckledy has the same plumage as the Cuckoo Marans. It is named after the speckled pattern of its dark brown eggs. Production is around 260-270 eggs.

*Blue Belle* A French hybrid, this is based on a Rhode Island Red x Cuckoo Maran crossing. Its plumage is an all-over grey (blue). Around 240 brown eggs are produced.

*Speckled Star* This comes from France and is a commercial strain of the traditional Cuckoo Marans. It lays around 200 dark brown eggs.

**Breeds for blue-green eggs**

The Araucana has a gene for blue-green egg shells, making the breed popular as a source of speciality eggs. The Cream Legbar is a traditional breed that was developed from strains of Buff Leghorn, Barred Plymouth Rock and Araucana. As the genetic factor for blue-green egg shells is dominant, it also shows up in the Cream Legbar's eggs.

The Araucana has a crest and as this is also dominant it shows up in the Cream Legbar and sometimes in the hybrids that are based on it. In recent years, these more productive hybrids have been developed to meet the demand for the unusually coloured eggs which now range from olive to blue, to green. They include *Old Cotswold Legbar* and several commercial Araucana crosses to which breeders give their own names .

## Problems in the Laying Flock

*Hen problems*

**Reluctance to go into house**: too hot and humid, or presence of red mites.
**Floor-laid eggs**: not enough nest box space or nest boxes badly placed.
**Egg-eating**: heat, lack of water or nutrients, or infrequent egg collection
**Feather pecking and aggression**: flock density too high, genetic pre-disposition to aggression, lack of nutrients, presence of external parasites, lack of 'head-height' browsing, boredom, mixing of different ages and sizes leading to 'pecking order' problems.
**Excessive feather loss**: Unbalanced feeding combined with too much time inside.
**Prolapse**: Unbalanced feeding at start of lay producing excessively large eggs.

*External egg problems*

**Egg with a band around the middle**: Egg has been temporarily halted in the hen's system, usually as a result of sudden shock or stress.
**Distorted, with soft ends or uneven surface**: if a regular occurence may be the result of infection. Get veterinary advice.
**Soft-shelled egg**: Sometimes found at the beginning or end of lay, or in old hens when system is in state of change. May also be the result of shock, stress or disease. If it continues, get veterinary advice.
**Rough shells**: Called thumbprint' or 'slab-sided' - uneven coating. Infection or age.
**Wind egg with no yolk**: Sometimes found in pullets commencing lay or in older hens finishing lay. May also be the result of shock.
**Small eggs**: Genetic tendency. Inadequate feeding. Too much light, too early. Also in old hens coming to the end of lay..
**Blood smear on shell**: Result of straining if egg is large.
**Pinprick blood spots on shell**: Presence of red mites.
**Pinprick black spots on shell**: Faeces of red mites.
**Chalky layer on shell**: Thin layer of extra calcium deposited as a result of egg being retained too long in system. Usually caused by shock or stress.
**Pale shell**: Hot weather resulting in reduced feed intake, lack of cool water, infection or red mite attack. If regular occurrence consult vet.
**Loss of colour in previous dark eggs**: Strong sunshine on hens' backs.
**Muddy or dirty eggs**: Mud brought into house and infrequent change of nest liners.

*Internal egg problems*

**Green yolk**: eating shepherd's purse or acorns on the pasture.
**Pale yolk**: normal in winter when grass is not growing actively.
**Fishy smell**: excess fishmeal in feed or egg contaminated by taint during storage.
**Onion or garlic taste**: wild onion and garlic in the pasture.
**Watery albumen**: ammonia in litter or infection. Consult vet.
**Moving air space**: ruptured egg membrane or shell damage.
**Over-large air space**: old egg
**Dark, irregular patches**: mould and fungal contamination.
**Fertile egg**: laying flocks should not have a male running with them.
**Internal blood spots**: blood escaping from hen's ovarian follicle. Can be the result of shock or infection. Consult vet if it continues.
**Meat spots in egg**: brown spots in the albumen from oviduct wall tissue. More usual in old hens and in some genetic strains.
**Double-yolked egg**: result of two yolks being released into oviduct at same time and enclosed in one shell. Fairly common in large eggs.

# Managing layers

If the pullets have been bred and raised by the producer, they may well have the same house for laying as they did when they were growing. This has several advantages. It gets them used to perching from an early age and stress occasioned by moving to new quarters is avoided. The only changes that may need to be made as they grow, is to provide larger feeders and drinkers. Nest boxes can be kept closed until required.

## Introducing pullets

Where pullets, either bought-in or raised elsewhere, are to be introduced to a laying house, the perches, feeders, drinkers and nest boxes should be in place and ready. Move the birds into the house gently and without rush for it is a stressful time for them. Evening is the best time to do this. Once they are all inside, close the pop-holes and keep them confined until the following day. This gives them time to settle down and establish where 'home' is. The following day they can be allowed out onto clean range that has not been used previously by poultry. Needless to say, the boundary perimeters should be walked every day, and the electric fencing checked to ensure that all is well.

## Daily management

Opening the pop-holes in the morning presents a good opportunity to observe the pullets as they emerge. Are they eager and ready to come out? Are there any that appear cowed and nervous? This first inspection, followed by several others during the course of the day, helps to identify problems such as aggression or potential illness. It also gets the chickens used to their handler, ultimately seeing him or her as 'head of the flock'.

Talking to the birds may seem eccentric but, once the pullets recognise their handler, the sound can have a calming effect on them. Many producers have also found that having a radio playing near the house has the same effect, as well as helping to keep foxes away.

The condition of the house interior should also be checked daily in case areas of the floor litter need to be raked, replaced or added to. A smell of ammonia should not be detectable. If it is, action is called for. Nest box liners should also be checked and if necessary, replaced.

Effective ventilation is essential to avoid lung and other infections. If the atmosphere feels damp and close, the humidity is too high and there is not enough air circulating. Open the vents and ensure that all the available pop-holes are fully opened. In some cases it may be necessary to open the main door as well. Ventilation is not usually a problem with small houses, but controlling air flow can be more difficult in larger ones where a fan may be necessary, particularly in very hot weather. Conversely, there may be too much draught entering so that air flow is more akin to a wind tunnel. It is here that mobile houses are so much more adaptable than large ones. They can be sited away from prevailing winds, and if there are pop-holes on both sides, the ones on the leeward sides can be opened while the others are closed.

Manually filled hopper feeders and drinkers should be attended to, while automatic systems are checked to ensure that they are delivering properly. Feed and water needs to be available all the time birds are awake, and ideally this will include inside and outside facilities for them. Outside feeders and drinkers should be in the shade, and shelter provision made available for the hens so that their sense of security is enhanced.

Pullets that have been on an organic grower's ration, can be switched to an organic layer's ration as they come to the point of lay at around 18 weeks.

## Lighting

If eggs are to be available in sufficient numbers during the winter, artificial light must be made available in order to extend the natural daylight hours. The standards require that this may only be used to prolong day length up to a total of 16 hours, and the day must end at dusk, so the artificial light is provided before dawn. This ensures that the birds go to roost when the natural dusk arrives and any eggs are laid earlier in the day.

Light systems are available for static and mobile houses, with the latter often using 12-volt battery systems. A pre-programmed timer is essential to ensure that the light comes on and goes off at the appropriate time. They need to be checked and adjusted regularly.

Artificial light can be used for laying ducks as well as chickens, and also to bring breeding birds to a point of lay when fertile eggs are required for incubation. It is important not to give artificial light too early to pullets that are due to lay for the first time. Free-range pullets need to have grown adequately to cope with extensive conditions when they are laying. If too much light is given, they may start laying before their physique can cope.

## Avoiding feather pecking problems

Most of the problems that are likely to be encountered are indicated in the table on Page 70 but a great deal can be done to avoid them in the first place. Feather and vent pecking that can ultimately lead to cannibalism is a major problem in some flocks. (Hybrid breeders are selecting for increased docility and this is a factor that should be taken into consideration when choosing a breed). Aggression is much more likely to occur where there is a high flock density. Mixing different age groups and sizes is also to be avoided, not only to minimise aggression but also to prevent disease transference. Where an incidence of aggression does occur, the culprit should be identified and removed. Keeping her in a pen for a day or two, where she can still be seen by

**Range of the Chicken's Natural Browsing Levels**

**Head level**:
(Pecking)
shrubs, trees, seeds, berries

**Intermediate level**:
(Pecking)
grasses, herbs, seeds

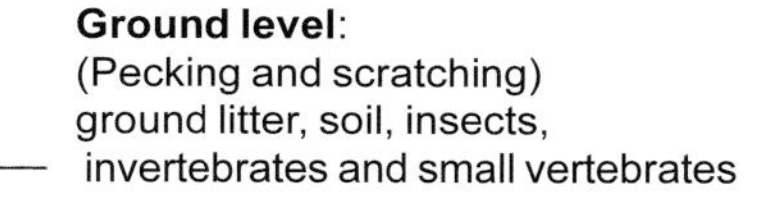


Chickens retain the behaviour of their Jungle Fowl ancestors in pecking and browsing at different levels. In the absence of suitable material at these levels they are more likely to peck at each other's feathers.

the flock will minimise the chances of her being attacked when she is returned. (Similarly, if a layer becomes broody and sits tight in the nest box, she should be removed and put in a cool pen to reduce her temperature, but where she can be seen, until she too can be returned to the flock). Owners of very large flocks might well consider such actions to be inappropriate in a commercial situation, and would merely dispose of the birds. To my mind, that is a factory farming approach, similar to the odious claim that only beak trimming can control aggression.

Beak trimming is inhumane and unnecessary if flock density is kept relatively low and a good level of management is maintained. Recent research has shown that the practice can affect the ability to feed, leading to food deprivation. (Ref: Prescott & Bonser. Journal of Applied Poultry Research. 13. 2004).

Bullying, feather and vent pecking, egg eating and cannibalism are far less likely to occur in small flocks that have plenty of room to roam about and are given the optimum conditions in terms of well-ventilated, mite-free housing, balanced feeding and plenty of shelter on the range. The availability of cold water at an optimum of 5°C in hot weather is also relevant, not only to avoid the problems referred to above, but also to maintain egg shell colour and quality. (Ref: *Shell Quality and Cooling Drinking Water*. Tankere, Bhandra & Dingle. Univ. of Queensland. 2001).

Another factor, in my view, is that 'pecking feedstuffs' should be made available at heights above that of normal pasture, to cater for an instinctive need to peck at 'jungle-height' shrubs. In their absence, the hens provide targets for each other. (Ref: Unpublished work: *Flock Observations. 1975-2005*. Katie Thear). Hanging up some greens at head height, or just above, can provide interest in different areas of shaded pasture and helps to prevent feather pecking. Seed blocks are also available that can be suspended above the house perches for the same reason.

Hanging up some greens can provide interest and helps to prevent feather pecking in the flock. Suspended seed blocks are also available for use in the house.

## Red mites

One of the most troublesome problems is the presence of red mites in the house. These can also trigger off feather pecking, as well as having a debilitating effect on the birds. If there is a sudden reluctance to go into the house, or an increased incidence of floor-laid eggs, their presence should be suspected. Pinpricks of blood on egg shells indicate where some have been squashed. Black pinpricks on the shells are their faeces. (See Health section).

## Egg collection and storage

Before collecting and handling eggs, the hands should be washed so handwashing facilities will be required near by. Where staff are involved, they should receive training in food hygiene and the avoidance of cross-contamination in relation to foodstuffs. Ideally, eggs should be collected twice a day, so that they can be taken for storage in a cool area as soon as possible. An ideal cool area is around 10-12$^{\circ}$C. If the temperature rises above 18$^{\circ}$C, a fan can be used to cool it. Eggs should be stored, pointed end down, in cardboard keyes trays, with plenty of air circulating around them, but placed out

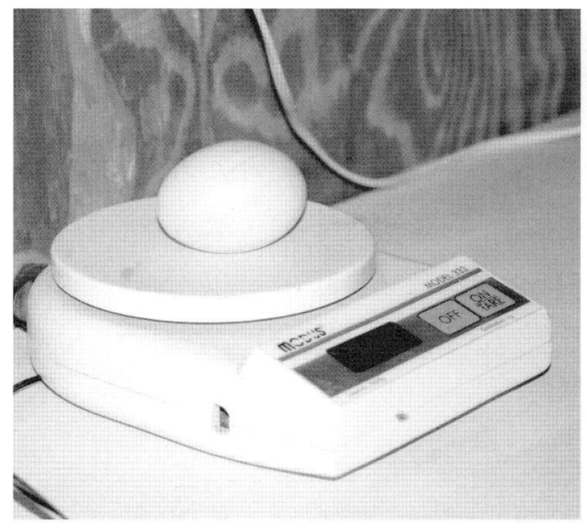


Electronic scales being used to grade an egg by size.

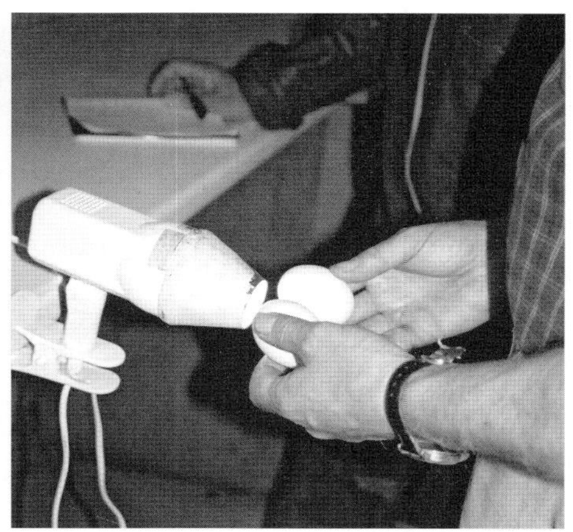

A candler beng used to check the size of the air cell and the internal quality of the egg.

of sunlight. If a distributor is used, the eggs will then be collected and taken to the registered packing station for grading and packaging. If the producer is registered as a packing station, grading and packaging will take place on site. Those unregistered, will be putting them in cartons to sell at the farmgate.

## Selling eggs

The eggs should be sold as soon as possible. Depending on the scale, selling will be by direct sales at the farmgate or through a distributor who will be registered as an 'egg packing station' under the Egg Marketing Regulations. In both cases, the producer must be registered with an organic certification body if the eggs are described as organic. However, the producer does not need to register with the Egg Marketing Inspectorate if the eggs are sold under the following conditions:

- there are no more than 350 chickens
- the eggs are the producer's own
- they are not graded and sold by size
- They are sold direct to customers at the producer's farmgate.

If a producer wants to sell eggs graded by size, or to supply shops, then he must register as an egg packing station. There is also a new requirement for all those selling graded eggs or selling eggs at public markets to use an identifying code for traceability purposes. This means stamping eggs with a code that is no less than 5mm in height, to include the following:

**0 UK 12345**

Production method - Organic

Country of origin

Producer's unique number

There is no minimum size required to register as a packing station, and it may be someone's home as long as the requirements are met. Full details are available from the local branch of the Egg Marketing Inspectorate. (See References). The basic requirements are:

- premises large enough to accommodate the work
- sufficient ventilation and lighting
- easily cleaned and disinfected
- suitably insulated to prevent eggs becoming excessively hot or cold
- no storage or handling of other products that might affect eggs in the area
- a candler to check internal quality of eggs and to measure the air cell
- a grader (scales) to weigh eggs
- a device for stamping eggs with a 'date of lay'.
- area and all equipment clean and free of smells that could taint eggs

## Egg grading

### Grade A

This is a designation that is applicable to fresh eggs. To use it as a description on an egg box, the following standards are required:

- no wet or dry cleaning of eggs
- free of smell
- normal, clean and undamaged cuticle (bloom)
- normal, clean and undamaged shell (no cracks)
- stationary air cell with height not exceeding 6mm (the older an egg, the larger the air cell)
- clear, limpid albumen, free of foreign bodies
- yolk, free of foreign bodies, visible as a shadow only without clearly discernible outline and not moving much away from centre on rotation.

### Egg size

The following are the recognised size grades that can be used as a description on an egg box:

| | |
|---|---|
| 73g and over - Very Large | 63g up to 73g - Large |
| 53g up to 63g - Medium | Under 53g - Small |

## Egg packaging

Eggs should be packaged in previously unused cartons. Registered producers should include a 'best before' date, which is a maximum of 28 days after lay. If a 'sell by' date is included, this should be 7 days prior to the 'best before' date. Also necessary to include is advice to customers to keep the eggs refrigerated until use. Eggs and cartons will also include the producer's code number for traceability purposes, as referred to earlier.

Unregistered producers who are selling only at the farmgate, may be using plain cartons or have a label with their name and address and an organic

Selling direct to callers at the farmgate or farm shop provides the best return.

logo. They may also wish to include a 'best before' date. Cartons, pre-printed and plain labels, and label dispensers are available from specialist suppliers.

It must be emphasised again, that although a producer may not be registered as a packing station, he must still be registered with one of the organic certification bodies if the description 'organic' is to be used on the carton.

For small producers the best returns are to sell direct at the farmgate or farm shop. Here, a regular and local clientele can soon be built up, and traceability of products is not an issue.

Selling at Farmers'Markets is also worth investigating for they are becoming increasingly popular. They are normally held at a local venue about once a month, at the same time, eg, the first Saturday of the month.

Providing eggs in organic 'boxes' is also a possibility, but the producer will need to advertise and make contact with potential customers well in advance, to build up a regular list.

Cotswold Gold chick. The sooner chicks can go outside the better for they are sturdier and less prone to illnesses in these conditions.

Cotswold Golds on range. Note how feeders and drinkers have been placed outside. Whenever a buzzard flew over, these chicks made a dash for the houses and were all inside in under a minute.

# Table Chickens

*Broilers are associated with intensive farming.*
*I prefer to call them table chickens.*
(Katie Thear. Lecture. 1997)

Until just after World War 2 most poultry farms had their own breeding flocks to produce both egg layers and table birds. As referred to earlier, a favourite cross was Rhode Island Red male with Light Sussex females. Most of the females became replacements for the egg laying flocks, while the males were raised for the table. It was common to caponise or castrate these males so that they put on a lot of weight, rather than running it off. At first, caponising was physical, followed later by chemical methods, but now thankfully, the process is no longer carried out.

Breeds such as the Rhode Island Red and Light Sussex were also raised independently, as table or laying birds, because it was still the time of dual-purpose breeds, rather than the highly developed hybrids that were subsequently bred for specific markets. My parents' choice was the Barred Plymouth Rock, with hens providing eggs while males were table birds.

Cornwall developed the Indian Game breed (also referred to as Cornish Game and White Cornish). Based on the Malay, Red Asil and Black-Breasted Red Old English Game, its wide-legged stance provided a large quantity of rather dense breast meat, but it was slow growing, with breeder flocks producing comparatively few eggs. However, when the males were crossed with females such as Sussex, Orpington or Dorking, the progeny were large, relatively quick-growing and with an abundance of less dense and succulent breast meat. An interesting fact, for those who may wish to experiment with producing their own table birds from traditional sources, is that the Transylvanian Naked Neck carries a gene for reduced abdominal fat. This becomes apparent when it is crossed with traditional heavy or table breeds.

## Housing

A mobile house is undoubtedly the best option for rearing table birds. It can be placed near natural woodland or hedgerow shade, while at the same time providing access to pasture. The Soil Association standards specify a maximum of 500 birds per house (although 1,000 is allowed under certain conditions). This compares with the gross UK Organic Standards requirement of a maximum of 4,800 birds. (Other housing requirements are detailed on Pages 31-32). The maximum number of table chickens per hectare (2.47 acres) of land is 2,500 under all the standards.

As referred to earlier in the book, commercial mobile houses are available from a number of suppliers. They are light, easily moved on skids and have good insulation. Plasticised or galvanised steel sheeting with an inner, sandwich layer of insulation is effective, and the smooth walls allow no hiding places for red mites.

Most existing poultry houses, sheds, barns or out-houses can also be adapted for table chickens, although in the case of static buildings, easy access to pasture must be available. The birds will not need nest boxes, but the question is often asked – should table birds be given perches? Such a question is academic where intensive production is concerned, because the birds are floor-reared inside from start to finish. Some producers claim that perches adversely affect the breast meat, but all chickens have an instinctive need to gain height and to perch at night (Ref: *Animal Welfare*. Olsson & Keeling. 2002).

I always provided straw bales for mine and this proved to be an acceptable alternative perch, as subsequently borne out by research findings. (Ref: Animal Welfare. Lambe. 1998). A recent pilot study, has also demonstrated the successful use of straw bales that were introduced to table birds from the age of 7 days. (Ref: *Pilot Study.* Food Animal Initiative. March 2004). Where normal perches are provided, they should be wide and rounded, and placed no higher than 30cm from the ground for ease of access by the heavier birds.

# Breeds

## Broiler breeds

It was the Cornish and the Barred Plymouth Rock that gave rise to the first of the modern, commercial breeds – the Cobb. Initially, a genetically dominant, White Plymouth Rock was developed to meet the demand for a white-feathered bird that did not have dark stub feathers on the carcase. This white male could be mated to different coloured females but the progeny would always be white. Then, a Cornish male was crossed with White Rock females and the progeny were subsequently selected for fast growth. An interesting aspect of this selection was the change in the frequency of pea combs. A pea comb is a small comb with triple ridges. Initially, nearly 100% of the birds had such a comb. However, fast growth is associated with single combs, and selection produced birds that were 90-100% single-combed. It was this line that produced the Cobb 500, a white-feathered table bird that has dominated the broiler industry.

Developed for intensive, indoor production, the Cobb is certainly a fast grower and has been known to grow too quickly for its legs to cope with its weight. The males are particularly prone because of their even larger size. In a free-range environment, where they are fed a lower protein ration, they

All chickens have an instinctive need to perch at night. Straw bales make good alternative perches.

are slower-growing, and exercise helps to develop and strengthen their leg muscles. The Ross 308 is also a white-feathered broiler. It was bred for indoors, but can adapt to free-range conditions, although the Sherwood White which is a Ross x Hubbard cross is slower-growing and therefore better suited to organic production.

## Free-range table breeds

In recent years, slower-growing hybrids have been developed specifically for the free-range sector and these are the ones favoured for organic production. The Hybro broiler from Euribrid, for example, is white-feathered and has a slower growing rate in its first four weeks of development, then accelerates after 28 days when its muscles have developed adequately. Other slow-growing white-feathered breeds, as well as brown-feathered ones, are also available. In recent times, producers have crossed Cornish Whites with Rhode Island Reds to produce yellow-skinned, white-feathered and slower-growing table crosses. It is appropriate to ask why slower-growing birds are preferable to the fast-growing broiler hybrids. There are three reasons:

- avoiding leg problems.
- a better textured meat with more flavour.
- the humanitarian aspect of the birds being able to be outside.

Under organic standards, the minimum slaughter age for table birds is 81 days if they are not from slow-growing strains, with at least a third of their lives outside. (The Soil Association requires two-thirds of their lives to be outside). Where slow-growing strains are concerned, there is no slaughter age restriction. In intensive houses, broilers are slaughtered so early that they are barely off heat in many cases. Outdoor flocks usually have access to woodland areas or to hedgerow strips where there are a range of wild plants and where insects can be caught. All these add to the flavour of the meat, and in the provision of extra minerals. Meat from very young birds tends to be very bland and relatively tasteless.

It was the French who pioneered the woodland method of managing free-range table birds, and also of introducing the designation *Label Rouge*, a description meaning that the birds had been raised outside and allowed to grow at a natural pace. The French must also be credited for developing coloured hybrids for the table, and in so doing, helping to provide a commercial role for the traditional breeds. In the long term the future of any breed is linked to its economic interest. The breeding firm *Sasso*, for example, has developed a range of traditional primary breeders that can be used to produce a wide range of coloured feather table birds to suit different markets. These include red, grey, black and other colours, catering for different areas that traditionally have had their own type of table bird. The breeders include Rhode Island Red, New Hampshire, Sussex, Plymouth Rock, Wyandotte, Gris d'Espagne, Marans Fauve, Marans Noir, Marans Gris, Faverolles, Ficchoise, Bourbonnais, Houdan and Taiwan Tou Dij (a Naked Neck type). However, these are not the table birds themselves, but the breeders that are used to produce them.

The way in which this is done is to cross these males with small, prolific hybrid hens that have a recessive gene so that the progeny always resemble the fathers in feather colouring, and so also resemble the traditional table birds. However, the genetic input of the mothers ensures that the progeny have short legs and plump breast meat. They grow well, but at a natural pace and have a good feed conversion ratio. Sasso's range of recessive females includes dwarf or compact strains, heavy, light and auto-sexing. They also have a good egg production rate, an essential requirement for the economic production of growing stock.

Hubbard-ISA has also developed colour feather table strains in a similar way. For example, the red-feathered ISA 457 is the result of a cross between the S44 male and the JA57 female. Also available in this country are JA 257 and 757, as well as Mastergris and Coloryield. Local names are often given to some strains of table birds. Examples are Devon White, Devon Bronze, Cotswold White, Cotswold Gold, Poulet Gaulois, etc.

**Table Poultry Breeds**

| *Traditional Utility Breeds* | *Broiler breeds* | *Free-range breeds* |
|---|---|---|
| Light Sussex | Cobb 500 | Euribrid Hybro |
| Cornish (Indian Game) | Ross 308 | Sherwood White |
| Rhode Island Red | | Hubbard ISA breeds |
| Plymouth Rock | | Sasso breeds |
| Wyandotte | | |
| Dorking | | |
| Marans | | |
| Faverolles | | |
| Houdan | | |

White and brown-feathered free-range breeds are available in the UK under a range of names depending on the supplier)

# Management

Unless the producer is breeding his own table birds, the chicks will need to be bought in before they are three days old, preferably from an organic breeder if there is one available.

Until the chicks are feathered they will need indoor protection with a suspended lamp to provide artificial heat. The protected area obviously needs to be quite secure from rodents.

Depending on individual circumstances, the chicks may be reared in a protected shed or barn and then transferred to their ranging house later, or they may go directly into the house at day-old, as the photograph on Page 49 illustrates. The area immediately around the suspended lamp can be confined with hardboard walls that are gradually extended as required. Dry wood shavings or clean chopped straw can be used as litter flooring. For the first couple of days, rolls of chick paper can be laid down to prevent slipping and leg damage. Dropping some chick crumbs onto the paper teaches those slow to learn, how to feed, for the sound and action imitates the action of the mother hen in dropping foodstuffs for her chicks.

Gas-powered lamps are frequently used as a heat source. If electrically powered ones are used, dull emitter lamps are best because they provide warmth rather than glaring light. The height can be adjusted according to their needs. If they are all huddled in a tight ball in the middle, they are too cold and the lamp should be lowered. If they are in a wide circle at the periphery, they are too hot and the lamp should be raised. In my experience, this is a far better approach than worrying about trying to achieve the environmental temperatures often quoted. As the chicks grow, the lamp is gradually raised until they are fully feathered, when it can be dispensed with. This is usually around three weeks, but in particularly cold spells, it may be appropriate to continue with the heat.

For the first 5-6 weeks, a starter ration of organic chick crumbs can be given ad-lib. Initially, this is best placed in shallow containers on the ground

A protected brooding area with small dish feeders, manually filled feed hoppers, automatically filled drinkers and gas-fired brooders for day-old table chicks. This kind of arrangement can be used for all types of poultry.

where it is easily accessible to the chicks. Water, either in manually-filled or automatic drinkers, should be available at all times. Once the chicks are familiar with where to find food, they can be given hoppers or other feeders. As they grow, they can be given an organic grower's ration for the remainder of their lives, or an organic finisher ration can be given for the last couple of weeks before slaughter. Any change in diet should be gradual, taking place over several days. Where feeds are prepared on the farm, organically sourced ingredients should be used, with a minimum 65% of cereals or cereal by-products. Insoluble grit must also be available.

The sooner the young birds can be allowed outside, the better, for the fresh air and exercise is beneficial to health. As mentioned earlier, electric poultry netting is effective in protecting them from foxes, while shade, access to a dust-bathing area and shelter from the wind is all-important if the young birds are to avoid a setback in their development.

On a small scale, where ground is made available on a regular basis and where damp areas are dealt with immediately, there is unlikely to be a coccidiosis or other health problem. Where litter such as wood shavings or chopped straw is used as flooring, it should be raked through regularly to introduce air and allow for rapid drying.

Young table birds off-heat and ready to go outside.

## Slaughter

The birds will be ready for slaughter when the producer decides, as long as it is not before 81 days if non slow-growing strains are used. (There is no time restriction on slow-growing strains). Free-range table chickens will normally be around 1.8 to 2.3kg (4 to 5lb) at ten weeks, but be aware that they take longer to grow in winter when more demands are made on the system. They will continue to grow if kept longer, but it is a matter of producing what customers or the market demands. Checking out demand and the availability of local farmers' markets and other outlets is crucial, otherwise there will be a lot of birds left to store in a large freezer. Although the weights given above are normally acceptable, it could be that larger birds are required at Christmas and Easter.

If there is a local slaughtering facility that is licensed for organic production, and there are at least 50 birds, it is probably worth having them slaughtered, plucked, eviscerated and dressed by them. Large organic poultry farms that have their own slaughter facilities will often process poultry from smaller organic producers in their areas. On a large scale, the birds will probably be grown on a contract basis and collected from the farm. Average liveweight at 84 days is around 2.5kg – 3.5kg (5.5lb – 7.7lb)

## Small-scale on-farm processing

The ideal situation is for birds to be slaughtered on site for they are then spared the undue stress of travel. Up to a maximum of 10,000 birds can be slaughtered on-farm and the site must be registered with the local Environmental Health department. The *Welfare of Animals (Slaughter or Killing) (WASK) Regulations 1995*, specify the humane considerations that need to be followed, for it is essential that no cruelty is involved. Slaughterers need to be licensed if an electric stunner is used, but not if killing is by neck dislocation or decapitation of birds that were raised on that site. The Humane Slaughter Association has some excellent publications, including *Practical Slaughter of Poultry*, which is essential reading for anyone with table chickens, ducks, geese, turkeys and guinea fowl. Training courses and practical tuition are also recommended for anyone involved with on-farm slaughter of poultry.

Before slaughter, every effort should be made to keep the birds unstressed. Food and water should continue to be given until around 6 hours before when food can be withdrawn, but water should still be available. This ensures that the gut is relatively empty.

There should be three areas available: a barn or similar place where birds are slaughtered and plucked, a cool hanging area or chiller at 4°C where the birds are hung for 3-4 days, and a processing area where evisceration and dressing take place. Offal must be safely disposed of and the Environmental Health department will advise on appropriate methods.

Feather plucking needs to be done while the carcases are still warm. It is advisable to have as many pairs of hands as possible to help with this, as well as with the subsequent evisceration. Small-scale feather plucking machines and other processing equipment are available from suppliers.

During evisceration and dressing, the carcases should be kept at no more than 4°C, with the aim being to process as quickly as possible. There is no substitute for practical demonstrations of these processes, and attendance at a training course is advisable.

The dressed bird is usually placed in a plastic bag with a label showing the following information:

- type of bird, eg, organically raised chicken
- weight
- name and address of the seller
- recommended storage conditions, eg, keep chilled at 4°C or less
- use-by date, eg, 7 days
- Organic certification code (of the appropriate certification organisation)

The logo of the certification organisation can also be included if required, although this is not compulsory.

# Ducks

*I can honestly say that I know of no livestock which can prove of more interest or give so much pleasure.*

(Reginald Appleyard. 1949)

Ducks are waterfowl so they understandably require water. In fact, it is a requirement of all the organic standards that a stream, pond or lake is available for them. The minimum depth is stated as being that which is sufficient to dip their heads in. That would need to be around 15cm (6in) but my own feeling is that a minimum of 30cm (12in) would be more appropriate, as long as there is a natural current in the water. This would enable them to get into the water and dabble as their nature demands. Dabbling ducks (a category to which all farmyard ducks belong) spend most of their time on the water surface, dipping their heads and occasionally submerging.

The water needs to be clean and well aerated. Stagnation and pollution can not only affect the health of the ducks, but also transmit pollutants to their eggs in the case of layers. There are two ways to achieve this: either to provide a constant flow, as in a stream, or to utilise a pond that is large enough for there to be adequate aeration from the surface. A good rule of thumb is that a pond surface area needs to be at least ten times its depth for natural aeration to be effective. It is also vital not to have too many ducks for the appropriate volume of water. Although their movements provide some natural aeration, they may also destroy the natural eco-system of the pond or lake. Little research has been done on the volumetric water requirements of ducks, but my own findings are as follows:

**Minimum water depth and maximum flock density for domestic ducks**

| | |
|---|---|
| Minimum depth to immerse head | 15cm (6") |
| Minimum depth to dabble | 30cm (12") |
| Minimum depth to avoid water freezing | 90cm (3ft) |
| Maximum flock density in flowing water | 8 ducks per 1sq.m |
| Maximum flock intensity in large expanse of still water | 4 ducks per 1sq.m |

(Ref: *Starting with Ducks*. Katie Thear. 2005)

UK Organic Standards specify no more than 2,220 ducks on one hectare (2.47 acres), while the Soil Association has a maximum of 2,000, but 'poaching' of the grass can be considerable where ducks are concerned, especially in the damper areas. This can be particularly notable around pond edges.

I always found it worth providing a wide, smooth, flag-stoned pathway or ramp leading into the pond. This not only controls their access (for they much prefer having a gradual incline leading into the water) but the area is

Here wooden slats are used under the drinker for these laying ducks.

prevented from becoming a quagmire. The use of wooden slats is also often used where water troughs and drinkers are concerned. Some suppliers sell portable, rigid ponds for shallow water, with a ball-valve control to ensure that fresh water is always replenished.

The probing bills of ducks can cause damage to the banks of ponds, lakes and streams. This can be prevented by using *Nicospan*, an edge liner with 'pockets' through which posts are hammered in. Natural vegetation soon grows through it and hides it.

A moveable pond with ball valve attachment to maintain level of water
*(Domestic Fowl Trust)*

Ducks are dabblers and need to be able to go onto water.

## Housing

The maximum number of ducks allowed per house is 4,000 under the UK Organic Standards. The Soil Association standards specify a maximum of 500 ducks, with 1,000 being allowed only under derogation which must be applied for. For meat poultry in fixed housing, the stocking density inside is ten birds with a maximum of 21kg liveweight per sq.m. In mobile houses that do not exceed 150 sq.m of floor space and remain open at night, the maximum flock density is 16 birds with a maximum of 30kg liveweight per sq.m. In my experience, ducks are much better kept in small flock densities, with around 30 per house being the optimim number.

The basic housing requirements are similar to those detailed for chickens, although perches and nest boxes will not be required for table ducks. Where laying ducks are concerned, nest boxes that are more specific to their needs are a good idea. (See Page 93).

Ventilation is a key factor for ducks for they are hardy, cold-climate birds, and there should be plenty of headroom for them. Outside is their natural environment and with their water-proofed plumage, they are well adapted to adverse weather conditions. However, they dislike strong winds and so shelter in the form of windbreaks is advisable. Shade from strong sunshine is also required.

Sterile Mule ducks that are crosses between the Barbary and Pekin. The slight crest and fleshy protuberances above the bill are indicative of their Muscovy heritage.

## Breeds

Traditionally, the Aylesbury and Pekin were the main table ducks, with Khaki Campbells and Indian Runners providing the eggs. As is the case with chickens, many of the pure breeds of domestic ducks have been bred for exhibition purposes rather than for production so their performance does not meet commercial requirements, unless they happen to come from a particularly good utility strain. Commercial hybrid strains are available, although there are very few organic sources of stock at the time of writing.

The brown Kortlang strains based on the Khaki Campbell are capable of laying up to 350 eggs. Cherry Valley has a white feathered egg layer, the Cherry Valley 2000, which is based on the White Runner, as well as a table strain, the Cherry Valley SM3. They supply breeding stock as well as ducks for rearing, so it is not a problem for producers to breed and rear their own organic stock.

Other breeds that prospective producers may come across are the Pennine which is a cross based on the Pekin and Aylesbury, and was developed in the North of England by Will Bradley and Thornbers. The Gressingham duck refers to a strain of wild Mallard raised for the table on the basis of its more 'gamey' flavour. At one time, the Whalesbury hybrid was also used. This was a cross between the Welsh Harlequin and Aylesbury, but these days, the Welsh Harlequin has been bred to be much lighter and pure strains of Aylesbury are hard to find.

Breeders in France have also developed commercial ducks. Gourmand Selection have used strains of the Muscovy to produce the Barbarie ST14, a compact grey-feathered duck for the Label Rouge free-range sector. They have also produced white and grey Mulard (sterile) table ducks that are Pekins crossed with Barbary Muscovy strains. Grimaud in France have also used Pekin and Muscovy lines to develop a range of ducks. Grimaud Option is a Pekin based strain that lays around 230 eggs, while Hytop Mule is a sterile table duck. They also have a range of Canedin Muscovy lines in different colours for raising or crossing. French strains are available in the UK

## Management

Day old ducklings will need protected brooder conditions similar to those provided for chicks. (See Pages 83-84). Chick paper can be made available on the floor until they have found their feet, so that slipping problems and hock damage are avoided. It is normal for them to be in a brooder house that is not subsequently used for the outdoor flock, but this will depend on the individual producer. Using the same house for both stages is certainly easier for the small flock than it is for larger numbers.

Once fully-feathered they should be allowed out as soon as possible, as long as the weather conditions are mild. Soil Association standards require them to have outside access for at least two-thirds of their lives, while the basic UK Organic standards specify a minimum of one third of their lives.

They should not be allowed access to swimming water until they are at least five weeks old and have been outside regularly for at least two weeks. Only the producer is in the position to judge this, as weather and other factors will have to be taken into consideration. In the absence of a mother duck, gradual access allows sufficient time for preening of the oil gland to waterproof the feathers. Earlier access can produce bedraggled ducklings that receive a severe setback in their development.

Ducks 'shovel' up their food so the provision of suitable shallow feed pans is vital. Compound pellet feeds are better than mash because powder can easily clog up their nostrils. They can take in whole grains, but as with all poultry, insoluble grit must be provided for them. There is small sized chick grit available that can be put out for them from the end of the first week.

An organic chick starter ration is suitable for ducks, followed by an organic grower ration. Non-organic chick starter feeds should be avoided because they may contain a coccidistat which is toxic to waterfowl. Some suppliers will make up organic rations specific to ducks. Young waterfowl need a fairly high level of niacin (Vitamin $B_3$) for adequate development.

An organic hen layer's ration of pellets is suitable for laying ducks.

**Suitable formulations for duck feeds**

| | *Starter crumbs* (hatch to 3 weeks) | *Grower pellets* (3 weeks on) | *Layer's pellets* (from point of lay) |
|---|---|---|---|
| Protein: | 19% | 15% | 17% |
| Oil: | 4.50% | 3.25% | 3.50% |
| Fibre: | 4.50% | 7% | 3.50% |

A suitable feeding regime would be similar to that for chickens, with compound feed available on an ad-lib basis and grain given in the afternoons. As referred to earlier, grain is a good way of getting the ducks into the house at night, for they tend to move as a flock. Ducks do not show the same predisposition as chickens to cluster around the house, but they are flock birds and where one goes, the others tend to follow.

Drinking water should be available close to the feeders. A pond should be regarded as a dabbling area, not a source of drinking water, although the ducks will obviously drink a certain amount while they are there.

Where Barbary ducks are concerned, it is as well to be aware that the Muscovy is not a dabbler It originated as a perching duck in South America, from a different genetic line to dabbling domestic ducks whose ancestor was the wild Mallard. It is much more like a goose in its behaviour, preferring to spend more time on pasture than on water. As a perching bird, it should also have access to perching facilities such as straw bales. Where young ones are reared, it may be necessary to put a net over the brooding area, to prevent them climbing out. The Muscovy has also retained the ability to fly, although once settled in a particular environment, it is less likely to attempt it. Heavy hybrid crosses are also less likely to fly. The clipping of wing feathers to restrict flight is prohibited under organic standards.

The minimum age at which ducks can be slaughtered is 49 days for Pekin types, 84 days for Muscovy types and 92 days for Mallard types. In practice, killing at 12-13 weeks is common for this not only avoids the pin feathers that can be difficult to remove, but also ensures that non-organically sourced stock will meet the required conversion period. The sequence of on-farm slaughter and the regulations that apply are outlined on Page 86. Dry plucking followed by waxing produces the best 'finish' and the ducks can be packaged, labelled and sold as detailed earlier. Average weight of dressed birds is 1.6kg – 2.2kg (3.5lb – 4.8lb).

Easily accessible nest boxes designed for ducks.

## Duck eggs

As referred to earlier, duck eggs do not come under the Egg Marketing Regulations, but they can be produced in similar conditions to those provided for laying hens, with some adaptation to the house. They are not specifically mentioned by the organic standards either, although there are housing and flock density requirements for ducks in general. (See Page 89).

One of the misconceptions about ducks is that they do not require nest boxes on the grounds that they lay eggs anywhere. Hens would lay eggs anywhere as well if they were not provided with nest boxes, but when they are provided, they will use them. The same is true for ducks.

Nest boxes for ducks need to be easily accessible, with no high 'step-over' to negotiate in order to get in. In fact, they do not even need a roof, as the photograph above indicates. The emphasis is on ease of entry and exit, with a front board just high enough to contain the nest box lining material.

It is usual to wash duck eggs in warm water at 46-49°C before they are offered for sale.

As referred to earlier, Indian Runners and Khaki Campbells are the traditional egg laying breeds of duck. There are also commercial strains that are based on Khaki Campbell, Pekin and White Runner, depending on the breeder. Examples are Kortlang Campbell and Cherry Valley 2000.

Geese are grazers and must have adequate pasture

Goose eggs can often be sold to craftspeople who buy them for decorating. This is from a Chinese goose and illustrates the difference in size between it and the large chicken egg.

# Geese

*They can be kept to advantage only where there are green commons.*
(*Cottage Economy*, William Cobbett. 1821)

The wild Greylag goose, *Anser anser*, of the western hemisphere is the ancestor of most of our domestic geese. It has relatively static populations within its range of distribution and also has the tendency to produce random mutations or sports that are pure white. It is not difficult to see therefore, that all the conditions were there for early domestication.

The wild Swan goose, *Anser cygnoides*, of the eastern hemisphere gave rise to the light Chinese and the heavier African geese, both notable for the distinct knob above the bill. The Chinese also produce more eggs than other breeds of geese.

Most of the traditional breeds of domestic geese are currently bred for exhibition purposes, and include Embden, Roman, Toulouse, Brecon Buff, Buff Back, Grey Back, West of England, Pilgrim, Pomeranian, Chinese and African. If they come from a good utility strain, they will have reasonable production rates, but most producers opt for commercial strains.

There is little point in keeping geese, however, unless plenty of good pasture is available, for grass is the mainstay of their diet. The maximum number of geese permitted per hectare (2.47 acres) is 670 under the UK Organic Standards (600 under Soil Association Standards), but this to my mind is excessive unless frequent rotation and resting of pasture is possible. Allowances also have to be made for permanent breeding stock, if the producer is breeding his own rather than buying day-olds.

## Commercial strains

Generally speaking, most commercial strains have been developed from one or more of the following: Embden, Roman, Chinese and Toulouse. The Chinese gives a higher egg production rate, while the Toulouse provides increased body weight. The Embden contributes a good body weight, and white feathers, with the latter also being found in the Roman.

### Legarths

These are white-feathered geese based on the Embden and developed in Denmark. They have long, broad breasts and have an average growth rate of 7.2kg in 14 weeks. They are the most common commercial strain here.

### Grey Landes

This is one of the most common commercial strains in France, but is not as popular in the UK, where white-feathered birds are preferred.

Hardy young goslings allowed out during the day but housed at night. These are Embden x Roman.

## Housing

The best form of housing is an open-fronted and high-roofed shed or barn that can, if necessary, be closed off at night with wire mesh doors to exclude predators. This ensures that there is adequate ventilation, for it must be remembered that geese are hardy birds that do not like to be confined. As long as there is adequate perimeter fencing to keep out foxes, the house can be left open all night, and this is the best option. It is a commonly held belief that a gander will see off a fox. This is quite untrue as I know to my cost.

Housing needs are very simple, and as long as adequate ventilation is provided, the only other necessity is a well littered floor, using wood shavings or chopped straw. This should be kept clean, raked and replaced as necessary to minimise the risk of disease. A maximum of two geese per 1 sq.m of floor area is the Soil Association's specification.

Drinking water should be available at all times, as well as access to a pond, lake or stream when they are outside, as referred to for ducks.

The basic standards allow up to a maximum of 2,500 geese per house. The Soil Association standards permit 250 geese per house, with a maximum ranging distance from the house of 100m. Under certain conditions, however, a maximum 1000 birds are allowed per house.

Geese, such as these Legarths, are flock birds and are easy to control.

## Management

The best option is to raise one's own stock, unless there is an organic source of day-old goslings nearby. A traditional ratio for breeding geese is two ganders for every nine geese, but it is important that they are able to sort out their own breeding sets in good time before breeding starts. It is not unknown for a gander to reject a particular goose. Having sufficient range area is also important at this stage so that there is no fighting. Once they have formed their breeding groups, there will not usually be a problem.

Geese are usually adept at brooding and rearing their own goslings, as long as they are in a protected area. They often choose the most unlikely places to build nests. I often found that the best solution was to stack straw bales around the broody on her nest. Rats and crows can be a menace so overhead netting may also be necessary.

Another problem is that geese often want to share the same nest so that eggs at different stages become mixed up and are left when the earlier ones hatch. Artificial incubation is often the answer, but there are problems here too. It is important for example, not to let the first gosling to hatch see you before it has seen its siblings, otherwise you will be imprinted as 'mother goose' and it will want to follow you everywhere.

## Rearing

Artificially hatched goslings, or those that have been bought in as day-olds, will need a protected brooding area with a heat lamp as detailed earlier for other poultry. A roll of chick paper (or another non-slip surface) on top of the wood shavings for the first couple of days helps them to find their feet and avoid leg problems such as 'straddle legs'.

A crumb starter ration should be made available to them in shallow containers. A suitable one that can be fed from hatch to 4-5 weeks is similar to the one shown for ducks on Page 92. Organic chick starter crumbs can be given as an alternative, but not that which is available for turkey poults because it is too high in protein. Over-feeding protein can lead to 'slipped wing' problems, where the wing either hangs down or projects at an angle because the muscles cannot support it. Strapping the wing to the body for a time can correct it, but only if this is done early enough, and regular checks are made to adjust the support. It is a problem that is best avoided.

It is also important to avoid non-organic chick crumbs for these will usually contain a coccidiostat that is toxic to waterfowl.

From around 4-5 weeks, an organic grower ration, such as that shown for table ducks can be given, while egg-producing geese such as Chinese, will do well on a pelleted ration of organic hen layer's feed. As with ducks, it is important to avoid dry powder feeds because geese can choke on them and their nostrils may also become blocked.

Drinking water needs to be available at all times. Indoor drinkers need to have restricted apertures such as wire guards to stop the goslings jumping in and getting wet, becoming chilled and fouling the water.

Apart from the danger of chilling, damp plumage can also invite other goslings to peck at the feathers. There is an early instinct to eat grass, and the wet, stringy feathers look rather like grass tufts to the goslings. I always found that cutting a turf of clean grass and putting it in their brooding area catered for this need and distracted their attention from feathers.

The sooner that goslings can go out onto an area of fine, fresh grasses, the better, as long as the weather is mild. Even at a week old, the exercise is good for their development and they are undoubtedly sturdier and less liable to infections. They will still need to be in a house at night and, depending on their age, may still require a heat lamp, although once they are feathered, it can be disposed of. From the age of 4-5 weeks, they will be hardy and only need to be in at night.

Access to swimming water should be prevented not only until the goslings are completely feathered, but also until they have been outside for a number of weeks. This allows time for feather preening to have waterproofed the feathers. Goslings that are naturally reared will be waterproofed at an earlier stage, thanks to the attention of the mother goose.

Fresh, clean pasture should be made available on a rotational basis, as detailed in the *Pasture* chapter. Over-used pasture, or one that has not been sufficiently rested, may have internal parasites such as gizzard worm that can be picked up by the goslings, leading to an emaciated condition known graphically as 'going light'.

If the grass is good quality, most of the geese's diet will be provided by this, with a compound feed or grain such as wheat being provided as a supplement. Some producers soak grain before making it available, but this is usually only necessary with certain breeds of ornamental wildfowl. As long as drinking water and insoluble grit are freely available, domestic geese have no trouble in taking a whole grain ration. Organically produced wheat is available from feed suppliers. As referred to earlier in the book, feeding grain is also a good way of enticing them to go inside, if necessary.

Table geese often have an organic finisher ration for the last 3-4 weeks before slaughter. If one formulated for geese is not available, an organic duck or organic turkey finisher ration is suitable.

Most table geese will be produced for Christmas so it is vital, not only to have explored the market beforehand, but also to plan the stages of killing, dressing and selling in good time. The regulations that apply to home slaughtering and processing are referred to on Page 86. Geese are generally more time-consuming to dry pluck, although small plucking machines are available. It may also be necessary to wax-finish the carcases to get rid of remaining stub and down feathers. Liveweights are around 4.0 - 8.0kg, depending on the sex, for ganders will be heavier than females. 'Long-legged' or uneviscerated geese should be hung in a cool room until ready for dressing.

## Goose eggs

There is a small market for goose eggs within the crafts sector. As they are for decorating, there is obviously no need for them to be organically produced, but any surplus eggs need not go to waste.

The Chinese lays more eggs than any other breed of goose, although they are slightly smaller than those of the heavier breeds.

A well-strawed area of a house provides a nesting area, but care must be taken to replace soiled nesting material on a regular basis. Any eggs that are taken for sale to craftspeople should be washed before sale.

Young Norfolk Black turkey out on range.

Turkeys appreciate having perches outside as well as inside. These are Kelly Bronze.

# Turkeys

*Beef, mutton and port, shred pies of the best,*
*Pig, veal, goose and capon, and turkey well drest.*
(Tusser. 1573)

Introduced to Europe from Central America by the Conquistadores of Spain, the turkey rapidly gained a reputation as a table bird worthy of any feast. It is now the bird of choice in most households at Christmas.

Already a large and quick-growing bird, commercial development of the turkey has produced such massive creatures that natural mating is no longer physically possible with many commercial strains. Artificial insemination has to be used. These strains, with their attendant humanitarian concerns, are associated with intensive farming, not with organic production.

Pure breeds of turkeys are available and these are often the choice of small breeders who are concerned with maintaining the genetic heritage of traditional breeds. They include Bronze, Cambridge Bronze, Norfolk Black, Buff, British White, Bourbon Red, Slate, Lavender and Cröllwitzer. *Turkey Club UK* represents their interests as exhibition and utilitarian birds.

In recent years, commercial strains have been developed for the free-range sector, with the Kelly Bronze proving to be the choice of most organic producers. This is slower-growing and more suitable for extensive conditions. The fact that it has black feathers with a metallic sheen also gives it a more traditional image that contrasts with the intensive association of the white-feathered birds. There are also slower-growing commercial strains of White and Black turkeys.

Slower-growing turkeys are available in large, medium and small strains, so it is possible to choose one or more of these, depending on the requirements of customers. Most buyers will be constrained by the size of their domestic ovens, while hotels and other caterers can cope with large birds. It is essential to establish what customers want and it is usual when taking pre-Christmas orders to note the weight range required.

**Performance Data for Different Strains**

| | *Super Mini Bronze* | *Roly Poly Bronze* | *Wrolstad Bronze* |
|---|---|---|---|
| 20 weeks | 5.97kg - 9.37kg * | 6.28kg - 9.87kg | 6.59kg - 10.36kg |
| 22 weeks | 6.32kg - 10.10kg | 7.16kg - 10.65kg | 8.00kg - 11.20kg |
| 24 weeks | 6.64kg - 10.90kg | 7.47kg - 11.65kg | 8.30kg - 12.49kg |

* Hen - Stag

(Ref: Kelly Turkey Farms)

# Housing

For rearing day-old poults, a brooding area such as that shown on Page 104, and as previously described for other poultry, is needed. The confining wall around the heated area is gradually widened as the young birds grow, just as the heater is raised as they become more hardy.

Once the poults are hardy they can go out on pasture with access to a house for night-time shelter. The essential requirements were covered in the *Housing* chapter. Open-fronted pole-barns or sheds, such as those used for geese are appropriate for turkeys.

The Soil Association Standards specify a minimum perch space of 40cm per bird, and a maximum of two birds per sq. m of floor area. There should be no more than 250 turkeys per house, although more are allowed under derogation if conditions are found to be suitable after inspection. (The UK Organic Standards specify a maximum of 2,500 per house, a density and number that is relevant to very few small producers). Turkeys can panic and easily damage their wings during mass movement within a house. Smothering and feather pecking are also greater risks where large numbers are concerned. On pasture a maximum of 1,000 turkeys per hectare (2.47 acres) is specified under UK Organic Standards while the Soil Association stipulates 800. Shelter and protection from predators are required.

# Management

## Breeding

The ideal from the small producer's point of view, is to have a breeding flock of turkeys and rear his own replacements. These may be from pure breeds or from using slow-growing commercial strains. These are available as fertile eggs or as A/H (as hatched) stock so there will be males and females.

A breeding ratio of 1 male to every 6-10 females is appropriate, depending on whether they are heavy or light breeds. A breeding flock can be run together, unless different pure breeds are in separate flocks.

Turkey hens normally start laying at 28-30 weeks with around 50 eggs being produced in the first year. The provision of artificial light for the breeders will stimulate egg production in the darker months of the year but it should be remembered that organic regulations stipulate a maxmum of 16 hours of light a day. The eggs are best incubated artificially, preferably within 10 days of lay before hatchability begins to decline. As with all eggs destined for incubation, they should be washed in warm water and an egg sanitant to minimise the risk of pathogens.

The salient points for incubating turkey eggs and those of other poultry are indicated opposite.

**Optimum Conditions for Incubation and Hatching**

*Egg Storage before incubation*: Temperature - 15 -18°C. Humidity - 75%
Ideally incubate before eggs are 10 days old. Normal room temperature for 24 hours before placing eggs in incubator. Site incubator at normal room temperature.

| | |
|---|---|
| *Chicken eggs* Incubation time: 21 days. | Day 1- Day 18: 37.5°C. Humidity: 52%<br>Day 18 - Hatch: 37.0°C. Humidity: 75% |
| *Duck eggs* Incubation time: 28 days. | Day 1- Day 25: 37.5°C. Humidity: 58%<br>Day 25 - Hatch: 37.0°C. Humidity: 75% |
| *Goose eggs* Incubation time: 31 days. | Day 1- Day 27: 37.5°C. Humidity: 55%<br>Day 27 - Hatch: 37.0°C. Humidity: 75% |
| *Turkey eggs* Incubation time: 28 days | Day 1 - Day 25: 37.5°C. Humidity: 55%<br>Day 25 - Hatch: 37.0°C. Humidity: 75% |
| *Guinea fowl eggs* Incubation: 28 days | Day 1 - Day 25: 37.5°C. Humidity: 55%<br>Day 25 - Hatch: 37.0°C. Humidity: 75% |

(Ref: *Incubation: A Guide to Hatching & Rearing*. Katie Thear. 2003)

## Rearing

In the brooding area, wood shavings litter is preferable to chopped straw. There is a danger of straw being eaten causing crop impaction. Once settled in the area, the poults can be given an organic starter ration. Unlike non-organic starter feeds, this does not contain a coccidiostat. Typical organic, compound turkey feeds are shown below:

| *Starter* | *Grower* | *Finisher* |
|---|---|---|
| Oil: 6.2% | Oil: 4.9% | Oil: 3.9% |
| Protein: 23.3% | Protein: 21.3% | Protein: 15.5% |
| Fibre: 4.3% | Fibre: 3.8% | Fibre: 3.9% |
| Ash: 7.1% | Ash: 6.2% | Ash: 4.9% |

The turkey crumbs can be made available from hatch to around 4-5 weeks, and approximately 175kg will be eaten by 100 poults. It is important to have drinkers that do not allow them to climb in. When first hatched, they can drown, while keeping surplus water off the wood shavings litter lessens the chances of disease amongst the birds.

From around the age of 4-5 weeks a grower ration can be substituted for the starter crumbs. Approximately 800kg per 100 birds will be consumed during the period until a finisher ration is given from the age of 14 weeks until slaughter. Approximately 1500kg of this will be needed per 100 birds. Please be aware that these are all approximations, for the amount of food eaten depends on a number of factors, including breed, strain, weather, environment and the availablity of supplementary feeds on range.

A brooding area for turkey poults. *(Maywick)*

As soon as the poults are fully-feathered and hardy they can go out on range. The basic standards require that turkeys spend at least a third of their lives with access to range, but the Soil Association specifies a minimum of two-thirds of their lives, bearing in mind that organic turkeys must not be slaughtered before the age of 140 days.

Inside and outside perching facilities are necessary, for turkeys are by nature perching woodland birds. Straw bales or purpose-made wooden stands can be provided for them.

Grain such as wheat can also be provided on the pasture, either on the grass or in a moveable feeder. If the turkeys are growing more quickly than is required, the compound feed can be reduced or stopped, with a grain only feed being provided. Obviously access to insoluble poultry grit is necessary at all times. Outside drinkers are also required.

Exercise on the pasture with every effort being made to ensure that the whole area is being used will help to avoid situations where the turkeys are growing too quickly. They are also great foragers and will glean a consider-

Buff x Bronze cross.

able amount of food from the pasture and its environs so that supplementary feed can be reduced.

Access to clean pasture that has not been used by chickens is essential in order to minimise the risk of Blackhead disease (Histomoniasis) which is a serious disease of turkeys. It can be transmitted by chickens without their being affected. Ensuring that pasture and inside litter are free of permanently wet areas will also prevent Coccidiosis protozoans.

## Slaughter

Reference has already been made to the fact that organic turkeys must not be slaughtered before the age of 140 days. Unless adequate planning has been made, it is all too easy to find that the turkeys are not at the right weights for the particular customers. Most of the time it is because the birds are reaching their target weights too early. Secondly, producing Christmas turkeys involves a lot of work in a comparatively short period of time. Adequate help, time and resources are vital.

Details of small scale, on-farm slaughtering and processing are given on Page 86.

Guinea fowl fit in well into an orchard and help to reduce the incidence of insect pests.

They are perching woodland birds and need conditions that will suit them.

# Guinea Fowl

*The African fowls are big, speckled, humpbacked and are called Meleagrides by the Greeks.*
(*Rerum Rusticarum*. Varro. AD.46)

Known as a delicacy since Classical times, the sharp-eyed Guinea fowl of West Africa was called Tudor turkey in the 16th century and has continued to appear on our tables ever since. It is still a minority demand when compared with chicken, but as a gourmet product it has shown signs of increase in recent years. As far as the organic sector is concerned, it is virtually unexploited. A useful sideline is the demand for its feathers for the fly-fishing industry.

The Pearl Guinea fowl, *Numida meleagris*, so-called for the white pearl dots on its purple-grey plumage, is the most common type, but there are other colour varieties including Lavender, Pied, and White. It has not been developed commercially to any great extent, although table strains from France are now available. The best approach for potential organic producers would be to buy some of this stock, as young birds or as fertile eggs, for breeding purposes, and then produce their own day-olds as required.

For breeding purposes, the optimum ratio is one male to 3 - 5 females running in a flock. The sexes are identical and can usually only be distinguished by the slightly smaller 'helmet' of the hen and their differing calls. (The male utters a single shriek while the female has a two-note call). They are not appropriate near close neighbours although they are good 'watch-dogs' on the farm.

Females will produce an average of 100 eggs a year and these are best incubated artificially for they are not renowned as broodies. In parts of Africa, where broody hens are used to incubate guinea fowl eggs, a common practice is to place a fertile hen egg in with the clutch. The resulting chick then acts as 'leader' to the keets, so that they are easier to control. (Ref: New Agriculturist. 2005).

Guinea fowl are popular in orchards where their insect-hunting activities are useful pest controls. In fact their ability to feed on a wide variety of plants and other foods has been demonstrated by research findings. In dry seasons they will even feed on succulent plants for sources of water and minerals. (Ref: Article by Njiforth, Hebou & Bodenkampa. African Journal of Ecology. March 1998).

To ensure year-round availability, artificial light is needed for breeding birds, while the number of batches of day-old keets will need to be worked out carefully to meet on-going demand.

## Housing

Most poultry houses are suitable for guinea fowl and the general requirements are listed on Pages 31-32. They need more perch space than chickens, however, with a required minimum of 20cm. The UK Organic Standards specify a maximum of 5,200 birds per house, while the Soil Association requires a maximum of 500, although up to 1,000 are permitted under derogation.

On pasture, no more than 2,500 guinea fowl per hectare (2.47 acres) are allowed, with access for at least a third of their lives. (The Soil Association requires access for at least two-thirds of their lives, with pasture being no more than 100m from the house).

## Management

Guinea fowl are easily panicked in a building and my own view is that no more than 50 birds should be housed together. When fear strikes a large number of birds, the tendency is mass flight to one corner of the house. This can cause injury, smothering and death. It explains why in some large houses, they place wire netting across the corners to provide breathing space when flight congestion takes place.

Guinea fowl are woodland birds and need access to shelter when they are outside. Outside perching areas are also popular, with tree branches providing a high vantage point from which to survey the area. Outside feeders and drinkers are also appropriate, even if they are also provided inside.

Day-old keets need a sheltered brooding area with heat lamp, similar to that provided for any young birds. Perches are recommended at an early age. If they are being reared in a house that they will not be using for the rest of their time, straw bales can be used as temporary perching areas. Once they are in their final house they will, of course have normal perches, like those provided for chickens.

The keets need a fairly high protein ration and an organic turkey starter ration is sometimes given to them. Alternatively, an organic chick starter ration is suitable for them, although it is lower in protein. All non-organic starter feeds should be avoided because the coccidiostats that they frequently contain are toxic to guinea fowl as well as to waterfowl.

From 4-5 weeks of age they can go over to an organic grower's ration such as that provided for table chickens. As with most young birds, the sooner they are able to go outside the better, for there will be less likelihood of feather-pecking and other behavioural problems. An organic finisher feed can be given to birds scheduled for the table, while an organic layer's or breeder's

Preening of the feathers is an important part of the routine of all poultry. This is a Lavender Guinea fowl.

ration can be made available for the breeding birds. An adult guinea fowl will eat approximately 113g (4oz) of food a day. As with all poultry, insoluble grit should be there for them to take as necessary.

Wheat is popular as a supplementary feed and is also useful as an incentive to get them into a house at night. It must be said, however, that guinea fowl are often reluctant to go inside. Open-fronted sheds or barns, similar to those used for geese, are appropriate so that they go in if and when they want to. Many will prefer to perch outside in the summer. This is fine, as long as they do not fall foul of foxes, although with their high perching instinct and sharp eyesight, they are much better adapted to protect themselves.

The minimum age at which organically reared guinea fowl can be slaughtered is 94 days. At this time their average liveweights will be around 1.1kg - 1.3kg (2.5lb - 3lb). See Pages 86 for further details for on-farm slaughtering and dressing.

## Guinea fowl feathers

There is a small market for guinea fowl feathers in the fly-fishing industry. The feathers from adult birds fetch a higher price than those from young birds, although both are saleable.

# Health

*Health or constitutional vigour is the first important factor to consider.*

(Herbert Howes. 1949)

The most important aspect of health in relation to organic flocks is the avoidance of problems, rather than dealing with them once they have appeared. This involves having healthy strains of birds that are suited to an outdoor environment. In fact, breeding one's own birds not only reduces the amount of stress to which birds are inevitably subject if travelling is involved, but also allows the development of natural immunity to the particular site.

A health plan should be drawn up in consultation with a veterinary surgeon, indicating the pattern of health maintenance and disease control measures that are to be followed. Such a plan is necessary as part of a farm's conversion to organic status. Recommended are the use of complementary or natural therapies such as homoeopathy and herbal treatments.

Vaccinations are permitted if there is a known disease risk such as Marek's and Gumboro disease on a farm or nearby land. The certification body should be consulted beforehand. Although routine, in-feed medications are banned, the use of antibiotics is permitted in clinical cases that require them, following veterinary advice. If a bird requires such treatment, it must be given on humanitarian grounds even if it means that it loses its organic staus. A separate area where it can be treated is also required.

After medication has ceased there is a withdrawal period for eggs and meat, during which time they cannot be sold. These are as follows:

| **UK Organic Standards** | **Soil Association Organic Standards** |
|---|---|
| Eggs: 7 days | Eggs: 14 days |
| Meat: 28 days | Meat: 56 days |

This book does not set out to cover all poultry illnesses, for a good veterinary book will give detailed information where required. It is again emphasised that a close cooperation with a veterinary surgeon is advisable where an organic poultry enterprise is being embarked upon. Newcastle disease and Avian influenza are diseases whose presence must be notified to the authorities.

The following are the conditions that are seen by organic producers as being the most significant:

**Feather pecking** (See also Pages 72-73). The best ways to avoid the habit are:

- Have a low flock density inside and out.
- Choose docile strains of birds.
- Ensure that young birds have perches from an early age.
- Encourage birds to spread out and use the whole pasture area.
- Provide outdoor shade and hang head-level greens outside.
- Provide hanging seed blocks above perches inside.
- Provide grain scratch feeds in different areas of the pasture.
- Maintain good ventilation and avoid bright light in the house.
- Make regular checks for external parasites.

- Ensure that the feed is balanced and sufficient.
- Provide cold water at all times.
- Identify and quarantine culprits until habit is forgotten.
- Be especially vigilant during the moulting period.

**Excessive feather loss** All the above-mentioned factors are relevant in ensuring that the moulting period, when feathers are lost to be replaced by new ones, does not last more than 2-3 weeks. When it lasts for a long period or when the birds look almost bald, there is something wrong. The main causes are:
- Pushing birds to lay large eggs by over-feeding amino-acids.
- Insufficient exercise by staying inside the house rather than ranging.
- Stress caused by too high a flock density or other 'fear' factors.
- External parasites.

**Coccidiosis** Coccidia are protozoans that get into the intestines. If birds look hunched up and miserable and there is blood in the droppings, it is likely to be coccidiosis. The organism can survive for long periods in buildings and on pasture. A vaccine is available where other control methods are ineffective.
- Avoid having damp areas in the litter and on the pasture.
- Clear areas of dirt and faecal build-up regularly.
- Clean feeders and drinkers regularly.
- Rotate pasture frequently.

**Blackhead** Also a protozoan parasite, *Histomonas meleagridis*, can be carried by chickens without harming them but can be a major problem for turkeys. Resistant strains of turkeys should be selected where possible and areas where other poultry have been should be avoided. A homoeopathic nosode is available.

**External parasites** Lice are external parasites that can be seen scuttling about on the skin. Having access to natural dustbaths helps to control them, but there are also products that are acceptable to the certification bodies. Red mites are more difficult to control because they inhabit cracks in the house, coming out at night to feed on the perching birds. (See Page 70). Again, the certification bodies will allow certain products to control them, but effective steam-cleaning, blow-torching or lime washing of houses at the end of the season is advisable.

**Internal parasites** Regular pasture rotation is the best way of avoiding internal worms, for their life cycle will be broken by having land lying fallow or growing a crop. Worms can be introduced by wild birds, however, and a careful watch should be maintained for gizzard worm in geese or gape worms in other poultry. Watch out for poor feather condition, hunched up appearance and sitting for long periods. The certification body should be consulted if there is a problem.

**Salmonella** The Lion Code requires that all laying hens are vaccinated against salmonella, but equally important is avoiding it by only having birds from regularly tested and monitored breeding flocks. Every effort should also be made to exclude vermin from the site and maintaining a good level of hygiene management in clearing away dirt and debris, and providing clean feeders and drinkers. It should be remembered that there is a 'due diligence' requirement for producers to protect their customers against food poisoning risks.

# Flock Replacement

*Beginning and end, they are the same.* (Traditional saying)

Birds need to be replaced on a regular basis. They may be reared on the producer's own site or bought in as day-olds. Whatever the situation, each batch needs to be prepared for and the housing and environment dealt with after it has gone. The following gives the sequence of events:

**Preparations** Once registered with an organic certification body and the conversion requirements are fulfilled, the market for finished birds or eggs needs to be established. After that, liase with a local supplier of organically reared birds (unless they are being bred on site) in order to ensure regular deliveries of day-olds. Talk to a local vet about any precautions that may be necessary and maintain regular contact in case of problems.

Ensure that the house and pasture with its protective fencing and shelter are all checked and ready. Provide feeders and drinkers and if these are automatically fed, ensure that they are delivering properly. Site drinkers over slatted areas if possible. Where stock is being bred or raised from day-old, have sufficient areas and equipment for incubation and protected brooding.

Work out how much feed will be needed for specific batches, then liase with a local supplier of organic feeds for regular deliveries. Ensure that there is adequate and rodent-proof storage available but avoid buying more feeds than necessary for they can 'go off' before use if stored for a long perid.

**Introducing or moving birds** Quiet handling will minimise stress. Brooder lamps where day-olds are being introduced should be on in good time to ensure that the environment is warm enough for them. Also ready will be wood shavings litter with non-slip strips, feeders and drinkers. (See Page 83). Older birds should be introduced to housing in the evening so that they have time to adjust to the house before going out in the morning. Each batch of birds should have its own number for identification and the keeping of records.

**Eggs on a continuous basis** At least two flocks of different ages will be required to ensure that there is no dip in egg production. As the first flock moves into the moult, for example, a second will be coming up to the point of lay. Neighbouring organic producers can also provide each other with eggs if there is a temporary shortfall.

**Culling laying flocks** The productivity of laying flocks gradually declines and it is usual for organic flocks to be replaced after 2 - 3 years. In some small flocks they may be kept for a longer period. Large flocks will usually be culled by contractors who take the birds away, while small ones will usually be dealt with by the producer. It must be done humanely and poultry awaiting culling must not be panicked or subjected to stress. There is a legal requirement for dead poultry to be incinerated rather than buried on site.

**Rearing one's own replacements** It can often be difficult to find good breeding males. Because of this shortage, males from non-organic sources can be introduced as long as they are then reared organically. A breeding flock running with males should be separate from laying flocks so that fertile eggs are not sold as eating eggs.

**Clearing and cleaning** All housing must be emptied between each batch of poultry, then properly cleaned and disinfected to prevent cross-infection and the build-up of disease-carrying organisms. The same applies to all equipment and utensils. Houses can be cleaned by lime washing, blow-torching or steam cleaning. Certification bodies will provide a list of acceptable cleaning and disinfecting products.

All runs must be left empty for at least two months to allow vegetation to grow back, and for health reasons. (The Soil Association Standards require a gap of at least nine months). In the case of poultry for meat production this period should be not less than two months per year, but the Soil Association also requires a gap of one year after two years in use. These requirements do not apply to small numbers that are free to roam throughout the day.

**Poultry manure** There is a requirement that no more than 170kg of nitrogen per hectare, per year, is produced as a result of manure being deposited by the birds. Where pasture flock density is high, the rotation will need to be frequent in order not to exceed this limit. The following indicates what the maximum number of birds per hectare is in relation to 170kg of nitrogen:

Laying hens: 260. Table birds (3.5 crops): 560. Male turkeys (13.5kg 2.1 crops per year): 120. Female turkeys (6.5kg 2.4 crops per year): 260

It is also important to prevent nitrogen leaching into water courses for pollution of this kind can lead to prosecution. The application of manure within 10 metres of ditches and watercourses and within 50 metres of wells and bore holes should be avoided. The best approach is to have a low flock density on pasture and where manure is removed from houses, to compost it with straw until is is well rotted. It can then be applied to areas of cultivation, but not to where poultry are to be kept.

**Predators** Reference has already been made to the importance of excluding foxes, roaming dogs and badgers from the site. Electric fencing that is frequently checked is essential, but poultry house doors should also be securely closed at night in case there is a successful incursion. Overhead netting may be necessary when young birds first go out, in case there are predatory birds such as buzzards. (These are protected species and must not be harmed).

Rats must also be excluded from houses. Avoid having stacks of timber, junk or other convenient hiding places around the houses. These can often conceal rat runs. Spilt food obviously attracts them, particularly in areas such as underneath slats where the poultry cannot reach. Baiting stations and traps can be set up, but must be properly placed and monitored.

# The Organic Enterprise

*It's not only new, it's organic!*

## Starting a new enterprise

A poultry enterprise has the same demands as any other business so it is sensible to take some independent financial advice, as for example, from a good accountant. Before beginning there are three steps needed. A start up business loan will not be available without these, or if the necessary documents have not been produced.

• **Research** This begins with the market. Estimate both the shape and size of the market now and in the future, as well as the activities of competitors. Also estimate the demand for produce now and in the future. Examine ways to meet demand and compete effectively, and remember to include the marketing costs in cost estimates.

• **A business plan** This is in two parts. The first is about you: your age, education, experience, personal means, property, liabilities and relevant business connections; in particular how your abilities match the requirements of the business. If there is already a business and this plan is being produced to raise money to diversify into organic production, the following is needed:

Three years audited accounts together with brief details of the existing business: when it was established, how it has evolved, its current structure and accounting system, borrowing history and current commitments.

The second part is about the business idea, detailing the practical aspects involved and showing how it can be implemented. Illustrate this account with photos or drawings, particularly if converting buildings or something similar. Include details of legal and planning requirements and restrictions and contingency provisions for setbacks of various sorts. Show what personnel if any are required, and if there is already an existing business, show how the new enterprise will fit in. Are there shared costs and savings? Will there be enough time and resources available?

• **A cash flow forecast** for the first three years. (See costings and returns).

## Applying for organic certification

Contact the Organic Conversion Information Service (OCIS) for information and advice, which may include an advisory visit. Secondly, choose one of the certification bodies and apply to them for an application form. There will be guidance notes to help with filling in the form. It includes providing personal details, those of the farm, current and proposed enterprises. (Here the *Business Plan* above will come in useful). Also required are details of the poultry, the conversion plan for achieving organic status, proposed soil management and crop rotations, with details of field use and so on. After the application form and fee are sent off, the certification body will arrange an inspection of the site.

## Costings

There are two kinds of costs: *fixed costs* or overheads and *variable* costs. The former include the depreciation of the initial business outlays, and the variable costs are the day to day running expenses. Add up the initial business costs, and work out their depreciation. The real cost of these assets is not the buying and using them but the cost of replacing them regularly. They are likely to include: housing, vehicles, machinery, equipment, furniture, office equipment, computer and software, legal fees, advertising and above all the initial purchase of birds.

Running or variable costs are what is needed month by month to keep the enterprise going. They are likely to include: feed (the biggest cost), heat, electricity, wood shavings, insurance, labour, veterinary, and many more. Feed can be shown as £/tonne and all other costs as £/bird.

## Returns

Selling eggs or meat direct to the consumer produces the best return. Using a distributor, producing to contract or selling to retailers, hotels and restaurants brings in a lower return, while selling to the wholesale market provides the lowest return of all. Estimate what proportions will be sold through these outlets and what income they will produce. From this a projected profit and loss account can be produced.

**Farmgate sales** The most profitable way to sell organic produce is at the farmgate, with customers coming to you. Before this, the marketing aspect will need to have been formulated and put into practice. In other words, potential customers will need to be told that the produce is available, and convinced that they should come and buy it. Local advertising is therefore essential. This can include sending out press releases to local newspapers and radio stations, placing advertisements in local media and distributing leaflets in the local area. Local organic and self-sufficiency groups can also be informed so that their members know that quality organic produce is available in their area.

**Farmers' Markets** Increasingly popular is the local farmer's market that is usually held once a month at a specific venue. The *National Association of Farmers Markets* has details of those held all over the country.

## Keeping records

Keeping records is a vital part of any business. In the case of an organic enterprise there are additional records that need to be kept in order to comply with the appropriate standards. They include:

- financial records, sales and purchases, VAT.
- veterinary records, including details of treatments, withdrawal times, etc.
- records that are specifically related to the organic nature of the enterprise showing the origin, nature and quantities of all materials that are bought-in. These include:
  - livestock, origin, status • feed • organic raw materials
  - additives • other input materials • processing aids and their use

The above information merely provides a basic overview and should not be regarded as comprehensive. Current prices are not included either for these soon go out of date. It should also be remembered that regulations and standards are subject to amendment.

# References

## Bibliography

*Compendium of UK Organic Standards.* DEFRA. 2005.
*Soil Association Standards for Organic Farming & Production.* Soil Association Certification Ltd. 2004.
*Free-Range Poultry.* 3rd Edition. Katie Thear. Whittet Books. 2002.
*Technical Bulletin: Managing Organic Laying Hens.* Soil Association. 2004.
*Technical Bulletin: Rearing Organic Poultry for Meat.* Soil Association. 2004.
*Organic Poultry Production.* N. Lampkin. Welsh Institute of Rural Studies. 1997
*Starting with Ducks*. Katie Thear. Broad Leys Publishing. 2005.
*Starting with Geese*. Katie Thear. Broad Leys Publishing. 2005.
*Incubation: A Guide to Hatching & Rearing*. Katie Thear. Broad Leys Publishing. 2005.
*The Chicken Health Handbook*. G. Damerow. Storey Books. 1994.

## Organisations and Sources of Advice

*DEFRA Helpline*: 01224 711 072 www.defra.gov.uk
*ACOS* (Advisory Committee on Organic Standards) Email: organic.standards@defra.gsi.gov.uk
*OCIS* (Organic Conversion Information Service). Tel: 01224 711 072.
England - Tel: 0117 922 7707
Scotland - Tel: 01224 711 072
Wales - Tel: 01970 622 100
Northern Ireland (livestock) - Tel: 028 9442 6752.

*Egg Marketing Inspectorate*
Northern Region - Email: j.sweeting@emi.defra.gsi.gov.uk
Midlands and Wales - Email: b.marper@emi.defra.gsi.gov.uk
South and East Region - Email: mark.s.jones@defra.gsi.gov.uk
Western Region - Email: j.hainsworth@emi.defra.gsi.gov.uk

*Elm Farm Research Centre.* Tel: 01488 658 298. www.efrc.com
*Freedom Food Ltd.* Tel: 08707 540 014. www.rspca.org.uk
*Humane Slaughter Association.* Tel: 01582 832919. www.hsa.org.uk
*National Association of Farmers' Markets.* Tel: 08454 588 420. www.farmersmarkets.net
*Northwest Organic Centre* (Myerscough College). Tel: 01995 642 206. www.nworganiccentre.org
*Organic Centre Wales* (Aberystwyth University). Tel: 01970 622 248. www.organic.aber.ac.uk
*Organic Egg Network.* Tel: 01570 480 785. Email: maesyronnen1@aol.com
*Organic South West*. Tel: 01579 371 147. Email: osw@soilassociation.org
*SAC Organic Helpline* (Scottish Agricultural College). Tel: 01224 711 072. www.sac.ac.uk
*The Wholesome Food Association.* Tel: 01237 441 118. www.wholesomefood.org
*Utility Poultry Breeders' Association.* Tel: 01926 420962.www.utilitypoultry.co.uk

*Yorkshire Organic Centre*. Tel: 01756 796222.
Email: info@yorkshireorganiccentre.org

## Certification Bodies

*Organic Farmers and Growers Ltd.* (Code UK2). Tel: 01743 440 512.
www.organicfarmers.uk.com
*Scottish Organic Producers' Association.* (Code UK3). Tel: 0131 335 6606.
www.sopa.org.uk
*Organic Food Federation.* (Code UK4). Tel: 01760 720 444. www.orgfoodfed.com
*Soil Association Certification Ltd.* (Code UK5). Tel: 0117 914 2406.
www.soilassociation.org
*Bio-Dynamic Agricultural Association.* (Code UK6). Tel: 01453 759501.
Email: bdaa@biodynamic.freeserve.co.uk
*Irish Organic Farmers' and Growers' Association.* (Code UK7). Tel: 00 353 506 32563. Email: iofga@eircom.ne
*Organic Trust Ltd.* (Code UK9). Tel: 00 353 185 30271. Email: organic@eircom.ie
*CMi Certification.* (Code UK10). Tel: 01993 885 651
Email: enquiries@cmicertification.com
*Quality Welsh Food Certification Ltd.* (Code UK13). Tel: 01970 636 688.
Email: mossj@wfsagri.net
*Ascisco Ltd.* (Code UK15). Tel: 0117 914 2406.
Email:Dpeace@soilassociation.org

## Suppliers

### Poultry

*Burcombe Organics.* (Organic chicks and ducklings). Tel: 01769 550330.
www.organic-chicks-poultry.com
*Cyril Bason.* (Black Rock, Calder Ranger, Speckledy layers. Ducklings). Tel: 01588 673204/5. www.cyril.bason.co.uk
*Greenacres Farm.* (Sasso and Ross table poultry). Tel: 01603 891092.
*Kelly Turkey Farms.* (Commercial free-range turkeys). Tel: 01245 223581.
www.kelly-turkeys.com
*Legbars of Broadway.* (Old Cotswold Legbars, Burford Brown layers). Tel: 01386 858007. www.legbarsofbroadway.co.uk
*Meadowsweet Poultry Services.* (Dutch, Belgian and French layers, table birds, organically reared hens). Tel: 08451 651532. www.meadowsweetpoultry.co.uk
*Muirfield Hatchery.* (Black Rock layers). Tel: 01577 840401.
www.theblackrock.co.uk
*Piggotts Poultry.* (Goslings, ducks, free-range layers & table). Tel: 01525 220944.
*S & T Poultry.* (Sasso table poultry, Maran Noire layers, Guinea Fowl, Barbary ducks). Tel: 01945 585618.

### Housing

*ARM.* Tel: 01889 575055. www.armbuildings.co.uk
*Cosikennels Ltd.* Tel: 01953 718294. www.cosiarcs.com
*Gardencraft.* Tel: 01766 513036. www.gcraft.co.uk
*Hodgson Timber Buildings.* Tel: 01833 650274.www.hodgsontimberbuildings.co.uk
*Jaques International.* Tel: 01568 708644. www.jaquesint.com

*Liberty Livestock Systems.* Tel: 01296 748842. www.liberty-livestock.co.uk
*McGregor Polytunnels.* Tel: 01962 772368. www.mcgregorpolytunnels.co.uk
*Newquip.* Tel: 01765 641000. www.newquip.co.uk
*NFP Ledbury.* Tel: 01531 631020. www.nfpledbury.co.uk
*Potters Poultry.* Tel: 01455 553234. www.potterspoultry.co.uk
*Rivers Animal Housing.* Tel: 01233 822555. www.riversanimalhousing.co.uk
*Smiths Sectional Buildings.* Tel: 0115 925 4722.
www.smithssectionalbuildings.co.uk

**Electric fencing**
*Electric Fencing Direct Ltd.* Tel: 01732 833976. www.electricfencing.co.uk
*G.A. & M. Strange.* Tel: 01225 891236.
*Kiwi Electric Fencing.* Tel: 01728 688005.
*Rappa Fencing.* Tel: 01264 810665. www.rappa.co.uk

**Organic Feeds**
*Alltech (UK) Ltd.* (Organic feed supplements). Tel: 01780 764512.
www.alltech.com
*Bowerings Animal Feeds Ltd.* Tel: 01278 458191.
*Hi-Peak Feeds.* Tel: 01142 480608. www.hipeak.co.uk
*HJ Lea Oakes.* Tel: 01270 782222. www.hjlea.com
*Lloyd Maunder Ltd.* Tel: 01884 820534. www.lloydmaunder.co.uk
*Small Holder Feeds.* Tel: 01362 822900. www.smallholderfeed.co.uk
*Vitrition.* Tel: 01423 324224. www.vitrition.co.uk
*W. H. Marriage & Sons Ltd.* Tel: 01245 612 000. www.marriagefeeds.co.uk

**Health**
*Ainsworth Homoeopathic Poultry Health Products.* Tel: 01271 342 077.
*Crossgates Farm Homoeopathic Products.* Tel: 01729 824 959.
*The Homoeopathic Pharmacy.* Tel: 01974 241 376.

**Miscellaneous**
*Ascott Smallholding Supplies Ltd.* Tel: 0845 130 6285. www.ascott.biz
*AXT-Electronic.* (Automatic pop-hole doors). Tel: 0049 3691 721 070.
www.axt-electronic.de
*Breckland International.* (Pecka-block seed blocks). Tel: 01760 756414.
*Danro Ltd.* (Egg boxes, labelling equipment). Tel: 01455 847061/2.
www.danroltd.co.uk
*Domestic Fowl Trust.* (Range of poultry and equipment). Tel: 01386 833 083.
www.domesticfowltrust.co.uk
*Fishers Woodcraft (*Signs). Tel: 01302 841 122. www.fisherswoodcraft.co.uk
*Hengrave Feeders Ltd.* Tel: 01284 704 803.
*Parkland Products.* (Free-range feeders). Tel: 01233 758 650.
www.parklandproducts.co.uk
*Quill Productions.* (Free-range feeders and drinkers). Tel: 01258 817261.
www.quillproductions.co.uk
*Rooster Booster.* (12volt lighting systems). Tel: 01963 34279.
www.roosterbooster.co.uk
*Solway Feeders Ltd.* (Slaughtering & plucking equipment) Tel: 01557 500 253.
www.solwayfeeders.com

# Index